"During the past decade, innovative researchers have applied principles of engineering and physics to cancer research, in an emerging field known as physical oncology. This multidisciplinary research effort has led to encouraging results towards a better understanding of different aspects of cancer biology and oncology, from quantitative understanding of tumor growth and progression to improved detection and the treatment of cancer. This book introduces and discusses the advances made at the interface of engineering/physical sciences and oncology and explores the new frontiers in this field. The chapters are well written and I am especially impressed by the approach to developing accurate mathematical predictions of tumor drug response in experiments and in patients. I think this book is a significant contribution towards the biomedical community's efforts to improve its quantitative understanding of the treatment of cancer. I hope the book will encourage more scientists to enter the field of physical oncology."

Sanjiv Sam Gambhir, MD, PhD
Stanford University School of Medicine

An Introduction to Physical Oncology

How Mechanistic Mathematical Modeling
Can Improve Cancer Therapy Outcomes

CHAPMAN & HALL/CRC
Mathematical and Computational Biology Series

Aims and scope:

This series aims to capture new developments and summarize what is known over the entire spectrum of mathematical and computational biology and medicine. It seeks to encourage the integration of mathematical, statistical, and computational methods into biology by publishing a broad range of textbooks, reference works, and handbooks. The titles included in the series are meant to appeal to students, researchers, and professionals in the mathematical, statistical and computational sciences, fundamental biology and bioengineering, as well as interdisciplinary researchers involved in the field. The inclusion of concrete examples and applications, and programming techniques and examples, is highly encouraged.

Series Editors

N. F. Britton
Department of Mathematical Sciences
University of Bath

Xihong Lin
Department of Biostatistics
Harvard University

Nicola Mulder
University of Cape Town
South Africa

Maria Victoria Schneider
European Bioinformatics Institute

Mona Singh
Department of Computer Science
Princeton University

Anna Tramontano
Department of Physics
University of Rome La Sapienza

Proposals for the series should be submitted to one of the series editors above or directly to:
CRC Press, Taylor & Francis Group
3 Park Square, Milton Park
Abingdon, Oxfordshire OX14 4RN
UK

Published Titles

Published Titles (continued)

An Introduction
to Physical Oncology

How Mechanistic Mathematical Modeling
Can Improve Cancer Therapy Outcomes

Vittorio Cristini

Eugene J. Koay

Zhihui Wang

With contributions by
Jason Fleming, Sofia Merajver, John Lowengrub, and others

CRC Press
Taylor & Francis Group
Boca Raton London New York

CRC Press is an imprint of the
Taylor & Francis Group, an **informa** business

CRC Press
Taylor & Francis Group
6000 Broken Sound Parkway NW, Suite 300
Boca Raton, FL 33487-2742

© 2017 by Taylor & Francis Group, LLC
CRC Press is an imprint of Taylor & Francis Group, an Informa business

No claim to original U.S. Government works

Printed on acid-free paper

International Standard Book Number-13: 978-1-4665-5134-3 (Hardback)

Library of Congress Cataloging-in-Publication Data

Names: Cristini, Vittorio, 1970- author. | Koay, Eugene, author. | Wang, Zhihui, 1976- author.
Title: An introduction to physical oncology : how mechanistic mathematical modeling can improve cancer therapy outcomes / Vittorio Cristini, Eugene Koay, Zhihui Wang.
Other titles: Chapman and Hall/CRC mathematical & computational biology series.
Description: Boca Raton, FL : CRC Press, [2017] | Series: Chapman & Hall/CRC mathematical and computational biology | Includes bibliographical references and index.
Identifiers: LCCN 2016047968| ISBN 9781466551343 (hardback : alk. paper) | ISBN 9781315374499 (ebook) | ISBN 9781466551367 (ebook) | ISBN 9781315356884 (ebook) | ISBN 9781315337821 (ebook)
Subjects: | MESH: Neoplasms--therapy | Models, Theoretical | Treatment Outcome | Medical Oncology--methods
Classification: LCC RC263 | NLM QZ 266 | DDC 616.99/4--dc23
LC record available at https://lccn.loc.gov/2016047968

Visit the Taylor & Francis Web site at
http://www.taylorandfrancis.com

and the CRC Press Web site at
http://www.crcpress.com

Printed and bound in the United States of America by
Edwards Brothers Malloy on sustainably sourced paper

Contents

List of Figures

Figure 4.6 Drug-loaded nanocarriers lead to cell kill enhancement over bolus delivery. (A) Time–evolution curves of chemotherapeutic efficacy f_{kill} (Equation 4.11) of nanocarriers releasing drug compared with the estimated efficacy (symbols) of conventional chemotherapy, for parameter values $r_b/L = 0.05$ (dashed curves, upward-pointing triangles), 0.1 (solid curves, diamonds), and 0.5 (dotted curves, downward-pointing triangles), paired with BVF = 0.005, 0.01, and 0.05, respectively. (B) Same as (A), but normalized to the corresponding bolus values of tumor kill, $f_{kill,\,bolus}$.

Figure 4.7 Effects of combinatorial change in BVF and r_b/L on f_{kill}.

Figure 4.8 Testing the efficacy of drug-loaded nanocarriers in mice. Comparison of fraction of tumor killed measured across three different treatment BALB/c mice groups ($n = 10$ per group) over a period of 17 days (from day 14 to day 31 after 4T1 tumor cell inoculation) showing a roughly threefold increase in kill from nanovectored drug vs. free drug. At each time point, tumor volume measurements from the three drug treatment groups were first normalized to the measurement from the control (PBS) group (no drug treatment), and then to the initial tumor volume for each group; f_{kill} was then calculated as (1 – normalized tumor volume).

Figure 4.9 Measurements of tumor volume. Four treatment groups: PBS (control), free DOX, 1.0 μm porous silicon particle loaded with chemotherapy drug (PSP/drug 1.0), and 2.6 μm porous silicon particle loaded with chemotherapy drug (PSP/drug 2.6). Data were measured on days 0, 3, 7, 11, 14, and 17 after first treatment.

Figure 5.1 Mechanistic model of tumor kill from chemotherapy in patients with CRC metastatic to liver (f_{kill}). (A) Determination of model parameters from histopathological measurements. The panel on the right is a segmented, computerized image of a histologic section shown on the left. (B) The model (bold solid line) describes heterogeneous response within human cancer (symbols: 50 data points from eight patients with CRC metastatic to liver). The x-axis represents a pathological surrogate for drug diffusion penetration distance. These results demonstrate significant correlation of local tumor kill with perfusive and diffusive transport properties of tissue. Note that Equation 5.12 really depends only on r_b/L (also see [158]), meaning that L is not an independent parameter, and thus obtaining a statistically insignificant p-value for L is irrelevant here.

Figure 5.2 Prospective, patient-specific model predictions match outcomes of fraction f_{kill} of cells killed by chemotherapy in a third cohort of patients with CRC metastatic to liver. (A) Testing of Equation 5.12 from posttreatment histopathology (coefficient of determination $R^2 = 0.79$). (B) Predictions of Equation 5.12 using BVF parameter calculated (Equation 5.14) from pretreatment contrast CT perfusion measurements (open circles, average relative error ≈ 15% between prediction and actual). Multiple measurements per patient indicated by the same symbol (A) and standard deviation (B, filled circles). Model input parameters r_b (radii of

List of Tables

Preface

Of these sciences the gate and the key is mathematics.... He who is ignorant of [mathematics] cannot know the other sciences.... And what is worse, men ignorant of this do not perceive their own ignorance, and therefore do not seek a remedy.

ROGER BACON
Opus Majus, 1267

Cancer is a complex, heterogeneous, multifactorial, and multistage disease [1]. Its growth, invasion, metastasis, response to treatment, and therapeutic resistance all depend on multiple genetic and environmental cues [2]. It is currently the second most common cause of death in the United States after heart disease [3] and is on track to become the leading cause of death in the United States by 2030 [4]. Despite all efforts to fight the disease, including significant investments in basic and clinical research, it continues to impact every segment of society, leaving oncologists with the challenge of providing treatment for a diverse population across a vast area. At the same time, treatment costs continue to increase, treatment becomes more complex, and the ratio of patients to physicians increases as a result of population growth and aging, among other factors. The field of physical oncology involves novel approaches to battling the disease through the application of physical sciences, in most cases combined with mathematical modeling, to study various aspects of cancer, such as tumor growth and invasion dynamics, transport (and ultimately delivery) of drugs across tissue, cell death and genetic mutation, and even patient treatment outcome.

PHYSICAL ONCOLOGY

A significant problem in the treatment of cancer patients lies in the lack of widespread understanding and application of first principles of physical science, which function at multiple scales in time and space within tumors. From the atomic and molecular scales (e.g., the transport of pharmaceutical molecules across human cells) to the macroscale (e.g., tumor mass growth within organs and the human body), the dynamics of drug transport (perfusion, diffusion, and other means) and the dynamics of tumor growth can be described, explained, and modeled mathematically according to the universal conservation laws of physics. These mathematical models can be coupled to biological behavior and concepts that are derived from experiments, tissue specimens, and patient imaging, and are therefore specific to a particular cancer type, although they may vary across patients and within each patient's tumor.

While the focus on the development of new drug molecules has led to improvements in how specific cancers can be fought through multiple chemotherapy delivery systems, the fundamental problem of drug resistance within the human body (consisting of many interacting body systems) is still the main cause of poor patient prognosis; for example, only 26% of breast cancer patients survive five years beyond the beginning of their treatment [5,6]. Chemotherapy fails more often than it succeeds because the drugs, while effective in laboratory petri dishes and sometimes in animal models, often cannot kill the cancer in live patients as effectively. Drug delivery encounters resistance to its transport through blood vessels, extracellular diffusion in the tissue and cells surrounding cancer cells (stroma), and cell internalization because of the distance and associated "barriers" through which the drugs must pass in tissue. Delivering drugs into tumors (in therapeutically adequate amounts, for an adequate period of time, and with efficacy enough to kill tumor cells and prevent future proliferation) is driven by pressure and concentration gradients. Compounding this phenomenon is the presence of hostile microenvironments within tumors and treatment-resistant cancer cells that pass on their genetic data through natural selection. Oncologists commonly begin treatment of cancer based on measurements from initial tests, while further treatment decisions depend on empirical evidence gained through trial and error. This approach unsatisfactorily meets an individual patient's needs at the initiation of and throughout the course of treatment. We suggest that a comprehensive, inclusive, and theoretical approach to treating and understanding cancer based on the cooperation between physicists, chemists, engineers, pathologists, and oncologists can and should be adopted at the start of treatment if we are to hope for significant improvements in patient outcome.

RESISTANCE TO CHEMOTHERAPY

Current treatment strategies in oncology are limited by multiple factors, especially drug resistance, which is twofold mechanistically. First, tissue within the body at many scales resists the transport of drugs, just as a complex microstructured physical medium may resist molecular transport. Second, tumors become resistant to drugs when their microenvironments change and as the cancer cells proliferate according to natural selection. More specifically, chemotherapy drug molecules encounter resistance as they are transported through the blood, across organ tissue, and into the tumor. There, drugs must be absorbed across cell membranes and internalized into the cell to perform their function in killing cancerous cells. These initial elements of resistance—distance across which the drugs must transport and tissue barriers—constitute a tumor's first line of defense against prescribed drugs, which kill not only the cancerous cells but also healthy cells in the surrounding tissue. Once exposed to the drug, those cancerous cells that do not die (owing to physical [e.g., lack of drug access] or biological [e.g., efflux pumps and acquired mutations] drug resistance) can proliferate and pass their genetic material on to newly generated cells, which may inherit resistance to the drug. Cancers may develop rapidly after initial exposure to chemotherapy because of this natural selection, where drug-resistant cells survive chemotherapy and are selected to produce the next generation. Another factor limiting the efficacy of chemotherapy is the danger of creating high toxicity levels within the patient. Since drug transport processes can be described by physical equations and associated parameters (of course under reasonable assumptions), predictive, mechanistic modeling can be instrumental in understanding how and why tumors offer resistance to chemotherapy. Mathematical models developed by the authors, applying essential parameters gained from histopathology and other imaging measurements in tumors, help explain and quantify the transport phenomena of chemotherapy drugs.

PATIENT-SPECIFIC STRATEGIES

Oncologists looking for the most successful patient outcome strive not to overprescribe chemotherapy to avoid creating overly toxic environments within the body. Prescribing the proper drug, delivery platform, dosage, duration, and frequency is usually performed according to established protocols derived from basic patient measurements, while ongoing treatment and strategy are primarily derived from empirical data and the physician's preexisting experiential knowledge. Physical oncology uses mathematical models to provide predictions of treatment outcome with acceptable accuracy based on biophysical equations where parameters are gained from individual tumors in specific patients. Being able to predict how a drug will physically interact with both normal and cancerous cells within the body, and taking into account the differences across patients (individualized treatment), can help alleviate the problem of toxicity. Measurements gathered from common imaging and diagnosis protocols, such as contrast-enhanced computed tomography scans, magnetic resonance imaging scans, and patient histopathology, can quantify those parameters necessary for understanding an individual patient's tumor [7–9]. For example, pathologists can measure (directly or indirectly) drug perfusion penetration distances across a tumor, the diameters of the blood vessels, and the blood volume fraction of the tumor to provide key parameters involved in the accurate predictive analysis of the amount of cancerous cells that will be killed in the tumor [7]. In fact, even the noninvasive process of performing contrast-enhanced computed tomography scans can provide patient-specific imaging "biomarkers" that are useful in designing patient-specific treatment strategies [8]. With this wealth of information, physicians may be better informed in decision making for the patient's health.

By collecting data from these common clinical tests (with the help of pathologists and radiologists), oncologists can develop patient-specific drug regimens per tumor, even when more than one tumor is present within the same organ. The authors' equations and mechanistic modeling approaches can provide predictions about many outcomes surrounding cancer treatment, including tumor growth, drug uptake, cell kill, and patient survival. Recently published results of studies on pancreatic cancer, colorectal cancer metastatic to liver, breast cancer, and glioblastoma provide evidence of the validity of the theories contained; this book summarizes the results of some of these studies and associated review and position papers recently published. Additionally, advances in *nanotechnology* [10,11] have been applied as a drug delivery method in these studies with encouraging results—in some cases with manyfold efficacy in killing cancer compared with conventional drug delivery.

PHYSICAL SCIENCES–ONCOLOGY CENTERS

In recent years, new research centers for the study of physical oncology have arisen all over the country with the mission of facilitating advances in the study of how first principles of physical science can be applied to the study of cancer. Physical Sciences–Oncology Centers (http://physics.cancer.gov/centers/), operated in part by research universities and the National Cancer Institute of the National Institutes of Health, engage in various aspects of this mission. Ongoing and recent work at Physical Sciences–Oncology Centers that pertains to the content of this book includes understanding the evolution of cancer resistance to chemotherapy (Princeton University), the study of physical mass transport processes in cancer (Methodist Hospital Research Institute, Houston), the modeling of cancer growth and response to chemotherapy (University of Southern California), and the study of cellular cancer genesis using mathematical modeling (H. Lee Moffitt Cancer Center).

The principal authors of this book, based at the University of Texas Health Science Center at Houston McGovern Medical School and the Brown Foundation Institute for Molecular Medicine, and at the MD Anderson Cancer Center—all within the Texas Medical Center in Houston—are contributing to physical oncology through novel approaches in mathematical modeling of tumor growth, drug uptake, and cellular death rates. These are paired with the use of common clinical tests such as contrast-enhanced computed tomography scans, magnetic resonance images, and histopathology, where new imaging-based biomarkers are discovered, which then effectively become input parameters needed to inform patient-specific mathematical models for predicting chemotherapy and other treatments' efficacy—even predictions of survival. Prediction accuracy has already been tested and confirmed in both "retrospective" and "prospective" patient studies in the form of significant correlations between model results and measures of patient outcomes. Such breakthroughs offer genuine hope for significant improvements in strategies for fighting cancer once the mathematical models have been refined to achieve satisfactory individual patient fidelity. Findings made in current research and reported in peer-reviewed publications suggest that clinical requirements for successfully implementing the science are already housed in most cancer treatment facilities; additionally, it is possible that the science can be applied to a wide variety of cancer types. In a brief summary, theoretical work is currently being applied to the study of colorectal cancer metastatic to the liver, breast cancer, pancreatic cancer, brain cancer, and lymphoma, with future "moon shot" studies underway designed to develop mechanistic strategies for modeling and treating these and other forms of cancer according to first principles of physical science.

TIMELINESS

Why are these advances timely? While both cancer incidence rates and investments in research and development are increasing, so is the cost of treatment; *inversely*, access to health providers decreases as the aging population and the number of cases increases. Our system demands greater efficiency, better patient outcome, and a clearer or transformed path for physicians in treating patients. Oncologists working together with pathologists, physicists, mathematicians, and engineers can develop a system for precise diagnosis and patient-specific drug therapy that can effectively treat cancer and cure patients while increasing efficiency. The hope is that this book will create more awareness of these advances, demonstrate how this "physics and cancer" approach can address some of the major questions and barriers in cancer research and treatment through related modeling work, and finally, provide enough evidence for the public to better understand this approach, so that in the years to come, we may start implementing systems that lead to better quality of life overall.

Hopefully, this book provides the reader inspiration for further inquiry into the application of physical science to oncology. Our primary goal is to shed light on the evidence supporting physical theories of cancer growth, metastasis, and treatment, so progress may be made in increasing interdisciplinary collaboration that aims to eradicate cancer. With enough cooperation in the areas of cancer diagnosis, treatment design, and drug delivery, we can work together to improve the lives of millions of people who face one of life's greatest challenges.

Acknowledgments

This work has been supported in part by the National Science Foundation (NSF) Grants DMS-1562068 (V.C., Z.W.), DMS-1716737 (V.C., Z.W.), the National Institutes of Health (NIH) Grants 1U01CA196403 (V.C., E.J.K., Z.W.), 1U01CA213759 (V.C., Z.W.), 1U54CA149196 (V.C., Z.W.), 1U54CA143837 (V.C.), 1U54CA151668 (V.C.), and 1U54CA143907 (V.C.); the Rochelle and Max Levit Chair in the Neurosciences (V.C.); the University of Texas System STARS Award (V.C.); the University of New Mexico Cancer Center Victor and Ruby Hansen Surface Professorship in Molecular Modeling of Cancer (V.C.); and the Houston Methodist Research Institute (V.C.).

We cordially acknowledge Scott McIndoo and Armin Day, who professionally edited the book before publication.

Authors

Vittorio Cristini is a University of Texas STAR Fellow, a Rochelle and Max Levit Chair in the Neurosciences, a professor and director of the Center for Precision Biomedicine at the Institute for Molecular Medicine, the University of Texas Health Science Center at Houston McGovern Medical School, an adjunct professor of Imaging Physics at the MD Anderson Cancer Center, and a full affiliate member of the Houston Methodist Research Institute. He earned his "Laurea Summa cum Laude" in nuclear engineering at the University of Rome "La Sapienza" and a PhD in chemical engineering at Yale University. Dr. Cristini has taught at the Universities of Minnesota, California, Texas, and New Mexico. He is author of more than 100 publications in peer-reviewed scholarly journals and the book *Multiscale Modeling of Cancer: An Integrated Experimental and Mathematical Approach*, Cambridge University Press (2010). Dr. Cristini was recently included in a list of 99 "Highly Cited" mathematicians and the "World's Most Influential Scientific Minds" by Thomson Reuters.

Eugene J. Koay is an assistant professor in the Division of Radiation Oncology at the University of Texas MD Anderson Cancer Center specializing in the field of biomedical engineering. Dr. Koay earned his MD and PhD degrees through a joint program between Baylor College of Medicine and the Rice University Department of Bioengineering. Dr. Koay completed his residency in radiation oncology at MD Anderson in 2014. He is a coauthor of multiple publications on the subject of cancer genesis and modeling of cancer, and is currently pursuing research into the use of quantitative imaging techniques to provide predictive biomarkers for patient-specific cancer treatment and early detection of cancer.

Zhihui Wang is an associate professor in the Institute for Molecular Medicine at the University of Texas Health Science Center at Houston McGovern Medical School, as well as an adjunct associate professor in the Department of Imaging Physics at the MD Anderson Cancer Center. He is a coauthor of more than 40 peer-reviewed publications on the topics of hybrid multiscale modeling of cancer growth and metastasis, cell signaling analysis, cross-scale drug target discovery, and biophysical modeling of drug transport. Dr. Wang earned his BS in mathematics and mechanics at Hunan University, China, and his ME and PhD in biomedical engineering at Niigata University, Japan. He held faculty positions at the Harvard Medical School, Massachusetts General Hospital, and the University of New Mexico before joining the University of Texas Health Science Center at Houston. His research has been fully supported by the National Institutes of Health and National Science Foundation.

Contributors

Joseph D. Butner, MS, holds a master of science in mechanical engineering and is currently a PhD candidate in biomedical engineering at the University of New Mexico. His research focuses on agent-based and continuum modeling of biological systems. His interests include the development of hybrid multiscale mathematical models of developmental biology, and tumor growth and treatment, with a focus on quantification of the contributions of phenotypic hierarchies and endocrine and paracrine signaling to system behavior, as well as the pharmacokinetics and pharmacodynamics of chemotherapy–tumor interaction.

Jason Fleming, MD, attended Vanderbilt University as an undergraduate and received his MD from the University of Tennessee Center for Health Sciences School of Medicine in 1990. In 1999, Dr. Fleming completed a surgical oncology fellowship at MD Anderson Cancer Center and joined the faculty of the University of Texas Southwestern, where he became associate professor with tenure in 2005. Dr. Fleming returned to MD Anderson Cancer Center in 2006 to join the Pancreas Cancer Program. His clinical and research interests are solely focused on pancreatic cancer. These include the molecular genetics of pancreatic cancer metastasis, the development of early detection methods, and the preclinical and clinical testing of novel treatments for this disease.

John Lowengrub, PhD, who holds a PhD in mathematics from New York University, is a Chancellor's Professor of Mathematics, Biomedical Engineering, and Chemical Engineering and Materials Science at the University of California, Irvine. Dr. Lowengrub is also a coleader of the Systems, Pathways and Targets Program at the Chao Family Comprehensive Cancer Center at the University of California, Irvine. His research focuses on the use of mathematical modeling to understand how misrelated feedback signaling, metabolic reprogramming, and interactions between a vascularized tumor and its microenvironment can drive tumor progression and morphology and dictate the optimal course of treatment.

Geoffrey V. Martin, MD, who completed bachelor's degrees in chemistry, math, and physics from Ohio State University and holds an MD from the University of Cincinnati, is a radiation oncology resident at the University of Texas MD Anderson Cancer Center. His research has focused on the development of mathematical models to describe oncologic processes and to predict cancer treatment outcomes, especially regarding the role of the immune system in cancer therapy. His interests include applying multiscale computational models to clinical cancer care to further help individualize oncology treatments and improve cancer outcomes.

Sofia D. Merajver, MD, PhD, a physician scientist who treats patients and is also studying basic mechanisms of cancer progression, received her PhD at the University of Maryland. After graduating from medical school at the University of Michigan, Ann Arbor, she continued her training in internal medicine and oncology. She has published more than 275 papers, and is scientific director of the Breast Oncology Program and director of the Breast and Ovarian Cancer Risk Evaluation Program at the University of Michigan Comprehensive Cancer Center. Dr. Merajver utilizes basic principles of physics, biology, and mathematics to study cancer progression and to design new therapies against the most aggressive forms of cancer.

Eman Simbawa, PhD, holds a PhD in mathematics from the University of Nottingham, United Kingdom. She is an assistant professor in the mathematics department of King Abdulaziz University. Her research focuses on numerical approximation of eigenvalues of self-adjoint operators, numerical methods of integral equations, and mathematical biology. She works in collaboration with the University of Texas and the Department of Internal Medicine at King Abdulaziz University to develop mathematical models of cancer growth and response to treatment and immunotherapy interaction with cancer.

Alejandra C. Ventura, PhD, who holds a PhD in physics from the University of Buenos Aires, Argentina, is an assistant professor at the University of Buenos Aires and group leader at the Institute for Physiology, Molecular Biology and Neurosciences in Argentina. Dr. Ventura trained at the University of Michigan as a postdoctoral researcher and at the Institut Non Lineaire de Nice, France, as a visiting scholar. Dr. Ventura's research focuses on information transfer at the cellular level using mathematical and physical modeling with systems biology tools, for understanding the correlation between the topology of signaling networks and their emerging behaviors under both normal and pathological conditions.

Definition of Technical Terms

Acidosis: An increased acidity in the blood and other body tissue, that is, an increased hydrogen ion concentration.

Agent-based modeling (ABM): A modeling approach where each cell is represented as a separate, often unique entity. Each may contribute to the model result individually, allowing for explicit modeling and study of cell-specific factors, such as cell–cell interactions.

Anaerobic glycolytic metabolism: Metabolism of glucose in the absence of oxygen. Anaerobic glycolytic metabolism is much less efficient than metabolism in the presence of plentiful oxygen supply, only producing as little as ~5% of the potential cellular energy contained in a glucose molecule.

Angiogenic growth factors: Molecules that signal for the formation of new blood vessels to grow out from the existing vasculature; this process is known as angiogenesis.

CA19-9: A blood-based biomarker for pancreatic cancer. It is the only biomarker approved by the Food and Drug Administration (FDA) for this disease. Multiple studies have shown this biomarker to have a poor positive predictive value in identifying pancreatic cancer. It is not present universally in all patients.

Capecitabine: A prodrug or a drug that must be enzymatically processed before becoming an active drug. The active drug is 5-fluorouracil (5-FU). 5-FU is a thymidylate synthase inhibitor, which inhibits *de novo* synthesis of DNA.

Chemotherapeutic particles: Nanoscale blood-delivered tumor treatments, including small molecules, nanoparticle chemotherapy delivery vessels, and nanoparticles with other chemotherapeutic applications.

Continuum modeling: A good choice for modeling tumor growth on larger scales, as it describes model variables as continuous fields mostly by means of ordinary differential equations and partial differential equations rather than working at the resolution of individual cells. However, it is difficult to use this approach to examine individual cell dynamics and discrete events.

CT: Computed tomography. An x-ray-based imaging technique used to visualize the internal structures of an object often to generate three-dimensional schematics from images.

Diffusion: The movement of molecules from regions containing higher molecular concentrations to regions containing lower concentrations.

Diffusion coefficient: A numerical constant that relates the molecules (e.g., chemotherapy drug) diffusing through a medium (e.g., the tumor tissue) to that medium based on the physical properties of both, thus providing a mathematical quantification to molecular flux through the medium.

Gemcitabine: A nucleoside analog drug in which the hydrogen atoms on the 2′ carbon of deoxycytidine are replaced by fluorine atoms. The triphosphate analog of gemcitabine replaces one of the building blocks of nucleic acids (cytidine) during DNA replication, thereby arresting tumor cell growth. This causes apoptosis. The drug also targets ribonucleotide reductase, which inhibits production of deoxyribonucleotides that are required for DNA replication and repair. This also results in apoptosis.

Histopathology: Microscopic examination of tissue to identify disease states.

Human equilibrative nucleoside transporter: A cellular membrane transporter that allows nucleosides into the cell. Approximately 60% of patients with pancreatic cancer have downregulation of this transporter, and this has been correlated with poor response to the nucleoside analog gemcitabine.

Hybrid modeling: A modeling approach that combines aspects of both discrete and continuum modeling to provide a more complete description of the tumor and its environment. In most current hybrid models, individual cells are treated discretely but interact with other chemical and mechanical continuum fields.

Immunohistochemistry (IHC): A process that uses stained antibodies to target, label, and identify certain cellular architectural features or cellular types based on proteins expressed within the cell sample.

***In vitro*:** Latin for "in glass"; refers to studies conducted on cell or tissue cultures isolated from a complete living organism or from their natural surroundings.

***In vivo*:** Latin for "in life"; refers to experiments conducted in living, multicell organisms.

Inflammation: A defense response in the body to harmful stimuli.

Interstitium: The space between cells or small gaps within tissues and organs.

Intravital imaging: A microscopy technique used to obtain high-resolution images from within a living multicell organism, often through a window implanted in the organism.

Lysis: The cellular membrane degradation and subsequent spilling out of the cellular contents as a result of nonprogrammed cell death.

Mass conservation: As per the laws of thermodynamics, mass can be neither created nor destroyed, and thus is conserved in a closed system.

Master equation: A mathematical function in closed form that can be expressed in terms of a finite number of "well-known" functions, for example, exponent, logarithm, and trigonometric functions.

Morphometric measurements: Measurements that quantify the external shape of an entity in two or three dimensions.

MRI: Magnetic resonance imaging. A noninvasive imaging technique that uses magnetic fields and radio waves to view inside an object.

Nanotechnology: Engineered devices with dimensions within the nanoscale. In the case of cancer treatment, nanotechnology commonly refers to nanoparticles, which may be functionalized to deliver targeted therapy to the tumor environment.

Necrotic cells: Cells that have experienced unprogrammed or premature death due to unforeseen circumstances, such as lack of nutrients or oxygen necessary for cell survival.

Nuclear grading: A process used in pathology to help determine the stage, or severity, of a cancer tumor. It can also provide information about cellular proliferation. Cancer stages are assigned a numerical value, with higher values being associated with poorer prognosis. Increased nuclear feature disruption or increased nuclear size is often indicative of higher-stage cancers.

Perfusion: The process of delivering blood or blood-borne agents through the vasculature into capillaries and organs.

Stroma: The interconnecting regions between cells, tissues, or organs. Plays a mechanical and nutritional supportive role. This is the milieu of tissue and cells that are not specifically malignant, but are thought to contribute to cancer cell growth, resistance to therapy, and metastasis. The role of the stroma in these processes is complex, as some have shown that stromal components make cancer cells more aggressive *in vitro*, while others have shown that stromal components restrict metastatic spread *in vivo*.

Transport oncophysics: An idea that describes how cancer can be thought of as a series of physical aberrations at different scales that affect the molecular, cellular, tissue, vascular, and organism scales. These physical abnormalities are intimately linked to the underlying biology of cancer and affect the delivery of drugs to the cancer cells. Nanotechnology could be used to overcome these physical aberrations to improve cancer outcomes.

Tumor volume fraction: The volume fraction of a tissue occupied by viable tumor. Solid tumor tissues consist of not only tumor cells, but also tumor-associated normal epithelial and stromal cells, immune cells, and vascular cells.

VEGF: Vascular endothelial growth factor. This is a key growth factor that cells produce to stimulate the formation of blood vessels. Cancer cells may express this also, and it helps them recruit blood vessels to allow them to grow and spread.

Viable rim of a solid tumor: The outermost region or layer of the tumor volume, which is characterized by a rapidly proliferating population. Much tumor growth or expansion occurs due to cellular proliferation in this region.

CHAPTER 1

What Should Be Modeled in Cancer

Milestones for Physical Models

With Alejandra C. Ventura and Sofia D. Merajver

T HE FIELD OF MATHEMATICAL oncology is exploding, with increasing efforts directed at laying down the foundation for the phenomena of cancer development, growth, metastases, and response to therapies. This involves working at multiple space and time scales. The relative scarcity of data renders many parametrical models challenging and limits their robustness. Concepts from engineering and nonlinear systems can be applied to cell signaling in the cancer cell, yielding useful models. Here, we discuss the concepts of modularity, retroactivity, and pathway integration, and delineate some of the outstanding questions that we believe should be prioritized for modeling in cancer.

1.1 INTRODUCTION

The first known documentation of the cancer disease is from Egyptian papyri written around 2000 BC. The cases described in the ancient texts were breast cancer, which, at the time, was treated surgically by cauterization with a hot object. Recognizing the enormous scientific and technical progress during the ensuing millennia, it is still true today that many cancers are controlled primarily by removal of the tumor mass, albeit by more exacting and humane methods. Indeed, to substantiate this assertion, we need not look further than modern early detection strategies in cancer. Current efforts at cancer screening are considered successful in general if the diagnosed cancer is small enough to be amenable to complete excision. With rare exceptions (such as lymphoma), solid tumors diagnosed at an "early" stage (~1 cm) are largely treated by surgery alone.

Let us consider for a moment the concept of "early" detection. This is a clinically useful empirical concept because randomized screening trials have shown that early detection of cancer improves survival, if appropriate treatments are available [12]. However, the concept of clinical early detection does not translate well at the tissue level of tumor development; an early-detected 1 cm human solid cancer contains ~10^9 cells (~1 g), with 10^{12} cells (~1 kg) being a lethal load of cancer in humans, 1000 times larger than the mass presumably detected early. Whereas we have made very substantial advances in the understanding of clonality, genetic alterations in individual cells, and the aberrant gene and protein expression patterns that occur in cancer, we still lack sufficient knowledge of the integration of signals that give rise to the abnormal behaviors that make untreated cancer a lethal disease. Additionally, there is a lack of effective therapies for eliminating cancer in an individual patient that are more efficacious when given alone than surgical removal of the primary tumor even with early tumor detection (at the scale of ~1 cm). Moreover, our knowledge deficit extends to a nearly complete lack of interventions (other than selective estrogen inhibitors, lifestyle, and avoidance of known carcinogens) that are effective even earlier in the path to clinical cancer, at the preclinical cancer stages or in primary prevention.

We define preclinical cancer stages as those detectable changes at the cellular and tissue levels (either by morphology, topology, protein expression, or other quantifiable readouts) that indicate dysregulation of signal integration that may predispose invasive cancer. Examples of different tissue conditions in this category include lobular carcinoma in situ (breast), atypical epithelial hyperplasia (any epithelial tissue), and atypical nevi (skin). As a result, in the United States cancer affects one in two men and one in three women in their lifetime, constituting a massive public health problem. The urgency to find new ways to understand the underlying mechanisms of cancer and to design entirely novel treatment is enormous.

To take an extreme example, glioblastoma multiforme (GBM), a cancer that almost never metastasizes, is one of the most lethal cancers, in large part because aggressive surgery cannot be performed without unacceptable morbidity, and we lack effective ways to eradicate it *in situ*, even when extremely small. At the other end of the metastatic potential spectrum, we fail to achieve good survival rates in tumors such as inflammatory breast cancer (IBC), which is known to be intrinsically metastatic from its inception [13]. IBC is treated with multimodality therapy, including aggressive combination chemotherapy, rather than surgery up front. In spite of frequent complete clinical responses to initial chemotherapy, IBC is the most lethal form of breast cancer, with current five-year survival rates at only 45% [14,15]. *These examples are meant to highlight the notion that deficits in current cancer therapy are pervasive and transcend whether or not a tumor is highly metastatic.* We postulate that new breakthroughs in our understanding of cancer are needed to overcome these fundamental challenges. For the last century, formal scientific studies in the biomedical sciences have yielded large datasets and complex catalogs of observations and classifications. For many highly heterogeneous diseases, such as diabetes and cardiovascular diseases, this phenomenological approach has been extremely successful in finding cures, effective ways to control the symptoms, and good strategies for prevention. Cancer, possibly due to its enormous molecular and cellular complexity, is not as amenable to be conquered in this manner. We pose that a better approach is to understand how information is transferred within a cell and integrated across cells in tissues, particularly understanding the coding, decoding, transfer, and translation of information in cancer, insights that may likely be gained through modeling approaches.

1.2 CELL SIGNALING

Despite half a century of detailed molecular signaling studies, there is no unified theory to account for transmission of signals in different cellular environments and to predict the manner in which such signals are integrated. Basic and translational research has focused predominantly on activation of a single signaling pathway using a specific ligand for a defined receptor, providing information only for the pathway in question, and typically focusing on a limited number of endpoints. A pathway-centric approach remains incomplete, however, because of the intricate cross talk among cell regulatory pathways [16]. Indeed, a given molecular component can be associated with or interact with multiple signaling, transcriptional regulation, metabolic, and cytoskeletal process pathways [17]. Pathways cannot properly be considered to operate in isolation, as an alteration of one pathway can lead directly (via protein–protein interactions) or indirectly (via transcriptional or translational influences) to changes in others. Cells receive multiple signaling inputs either simultaneously or sequentially, with each signal varying in intensity and duration. Given the dynamic profile of the signal, being itself a packet of information the cell must integrate, there is much to be done to understand the integration of all the possible types of interactions between pathways.

Given the enormity of the unknowns, what is necessary to begin unraveling the problem? Successful identification of transmembrane receptors, intracellular signaling proteins, and transcription factors that mediate the responses of cells to intra- and extracellular ligands has generated a wealth of information about the biochemistry of signal transduction [18]. It is important to note that most biochemical and molecular biology experiments tend to be biased toward ascertaining large static differences between the expression (or modification) of proteins or genes, rather than subtle steady-state differences or significant dynamical profiles. In this manner, our current thought in signaling is driven yet limited by those types of data. Moreover, accumulation of molecular detail does not automatically yield improved understanding of the ways in which signaling circuits process complementary and opposing inputs to control diverse physiological responses. For this, network-level perspectives are required [19]. When the number of species in the network is large, parameter estimation becomes very challenging [20]. A plausible, alternative approach is to depict the pathway as a collection of modules that are connected with each other through input–output properties. Another useful approach when parameter estimation is challenging is to exhaustively explore the parameter space.

1.2.1 Parameter Space Exploration

A major goal for exploring the parameter space is to understand the uncertainty in a model's output that derives from a lack of knowledge about the exact value for a parameter that is assumed to be constant under certain conditions throughout model analysis. It is then possible to make inferences on the contribution of individual parameters to specific components of the system steady-state or dynamical properties.

For a given mathematical model of a biological system or process, the properties that the model is required to reproduce are first mathematically defined. A sampling method will next be used to search for these properties throughout the parameter space. Model parameter values will undergo statistical analysis to test whether a particular parameter is biased toward a certain value (or certain range of values) for the model to produce the targeted dynamics or steady-state properties. After this is done for all parameters, the results can be compiled to identify recurrent parameter values and any patterns that may form.

As a commonly used sampling method, Latin hypercube sampling (LHS) uniformly samples the values of parameters on a logarithmic scale, defining a certain range of each group, for example, catalytic rate constants, Michaelis–Menten constants, and concentrations. LHS is a statistical method for generating a sample of plausible collections of parameter values from a multidimensional distribution. In the context of statistical sampling, a square grid containing sample positions is a Latin square if (and only if) there is only one sample in each row and each column. A Latin hypercube is the generalization of this concept to an arbitrary number of dimensions, where each sample is the only one in each axis-aligned hyperplane containing it. In two dimensions, the difference between random sampling and LHS is illustrated in Figure 1.1. In random sampling, new sample points are generated without taking into account the previously generated sample points; one does not necessarily need to know beforehand how many sample points are needed (Figure 1.1A). However, in LHS, one must first decide how many sample points to use, and for each sample point, remember in which row and column the sample point was taken (Figure 1.1B).

For each parameter set, a model is numerically simulated, and those that have generated the desired outcomes are identified. A systematic approach to determine how model parameters contribute to system output is presented in [21]. Briefly, an enrichment test to find patterns in the values of globally profiled parameters with which a model can produce the required system features is performed. This is followed by a statistical test to elucidate the association between individual parameters and different parts of the system's features. A review of other approaches can be found elsewhere [22].

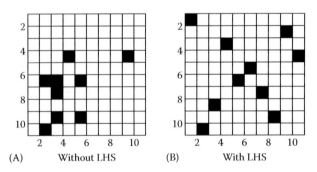

FIGURE 1.1 Difference between sampling methods in two dimensions. (A) Random sampling. (B) Latin hypercube sampling.

1.2.2 Modularity and Coarse Graining

Recent discussions of modularity in development [23] have emphasized that (1) a module is a biological entity (a structure, process, or pathway) characterized by strong and intricate primarily internal (rather than external) integration; (2) modules are individual units that can be delineated from their surroundings or context, and whose behavior or function reflects the integration of their parts, not simply their arithmetical sum; and (3) the definition of modularity works across multiple levels of the biological hierarchy, from subgenomic units to ecosystems. The components of each module interact with each other to form a coherent (larger) module, giving rise to a nest or conglomerate of modular structures. In a simplistic synthesis, modules can be insulated from or connected to each other. Insulation allows the modules to carry out many diverse functions without cross talk, whereas connectivity allows one function to influence another. Modularity plays a fundamental role in the prediction of the overall behavior of a system from the behavior of its components, based on the basic and preliminary assumption that the properties of individual components do not change upon interconnection.

Other researchers have refined and extended these ideas [24–26]. A fundamental issue that arises when interconnecting subsystems is how the process of transmitting a signal to a "downstream" component affects the dynamic state of the sending component. Several approaches have been proposed to extend ideas from electrical circuit design and related fields to analyze similar effects in the steady states of biological networks [27–29]. The same question can be considered in the more general and powerful context of what is called dynamical modules. This term encompasses any subsystem whose dynamical behavior is not significantly affected by its interconnections with other subsystems, whether because these connections are weak or because the subsystem topology effectively rejects even strong effects from the rest of the network [30,31].

What Should Be Modeled? The studies of dynamical modules should be distinguished from previous work based on the emphasis on global dynamical behavior rather than on local perturbations about a fixed point. This focus on functional biological dynamics also separates this perspective from that of the many researchers who have looked exclusively to network connectivity to define modules that are weakly linked to the rest of the system [16,24,32–36].

The appeal of obtaining simplified descriptions of network dynamics by replacing subnetworks with lumped elements has been apparent for some time, and several *ad hoc* approaches in specific systems have proven quite useful [37,38]. Importantly, it has been argued that global system behavior depends primarily on the bifurcation diagrams of subsystems, but not on their other dynamical properties [39,40]. Attempts to view the coarse-graining problem in a broader context have typically limited themselves by reducing differential equations to Boolean models [41–43] or to modules with other special properties [44,45]. Related work on multiscale modeling [46–48] and on automatic model reduction techniques [49–52] has some of the generality and mathematical rigor that needs to be attained, but without a similar concern for identifying modules that have biological function and can be interpreted by humans. The success of these earlier efforts indicates that the time is ripe for these new approaches. These will differ from previous attempts at modularization in taking a systematic and mathematically principled approach that can be generalized to many different systems, while at the same time yielding coarse-grained models that are not just computational tools but aids to human understanding. We are optimistic about the robustness of this methodology for future modeling in cancer.

1.2.3 Retroactivity, Substrate Competition, and Insulation

The "input–output" notion of signal propagation abstraction hides the fundamental mechanisms underlying information transmission, which may be crucial for understanding the causes of aberrant signaling. Mathematical and experimental studies have revealed that the biophysical mechanisms that lead to information transmission from an upstream module to a downstream module inevitably change the state of the upstream module [31,53–58]. These changes have been called retroactivity [31]. Retroactivity can cause significant cross talk between different signaling pathways that converge onto the same genetic or protein targets or onto the same signaling components [59]. Cross talk can also lead to long-lasting, undesired overactivity of signaling components, as commonly found in key regulators of growth in many types of cancer.

Nature has evolved a number of mechanisms to curtail cross talk between pathways. These mechanisms include mutual inhibition of pathways, kinetic insulation, and scaffolding [60]. What mechanisms (if any) has nature evolved to prevent cross talk due to retroactivity? It has been shown that covalent modification cycles can theoretically work in parameter operating regions to behave as insulators. That is, they can attenuate the effects of retroactivity and enforce unidirectional signal propagation [31]. These cycles implement negative feedback coupled with large input amplification to obtain the insulation property. In the case of a phosphorylation and dephosphorylation cycle, amplification is produced by the phosphorylation reaction, while feedback is produced by the dephosphorylation reaction.

Traditionally, one thinks of a cascade of molecular events as initiated by the recognition of a stimulus, in which chemical alterations at each stage lead to the activation of molecules that can in turn act on their downstream effectors. The mitogen-activated protein (MAP) kinase pathway is a typical example. We have shown, however, that because binding to one substrate makes an effector unavailable to other substrates and perhaps to the enzymes (e.g., kinases and phosphatases) that regulate it, such cascades can in fact support bidirectional signal propagation, and even in certain circumstances bistability and oscillations, in the absence of any explicit feedback mechanisms [58]. The substrate competition effect also mediates indirect interactions between different targets of a given enzyme. We hypothesize that subtle, indirect interactions of this sort, which can be viewed as a failure of insulation and modularity, may be an important cause of aberrant signaling in cancer; inappropriate cross talk of this sort has already been implicated in several cancer types [61].

What Should Be Modeled? We propose that modelers should build simplified descriptions of large, complex signaling networks in which collections of species and reactions are replaced by dynamically well-defined modules. These coarse-grained models will speed up simulations and computational parameter searches, and more importantly, they will serve as invaluable aids to mathematical intuition, analysis, and understanding. By clarifying the function of intricate reaction topologies and unveiling indirect regulatory connections, they will provide new insight on how information flows through signaling networks. In particular, they will make it easier to see a network's capacity for novel and unexpected behavior, thus modeling the network model in a truly predictive fashion.

1.2.4 Identification and Characterization of Common Dynamics Modules

Great efforts have been focusing in recent years on identifying network motifs, that is, agreed-upon modules [62], including bistable and ultrasensitive switches [63]; delay elements; kinetic proofreading [64]; networks that can achieve biochemical adaptation [65] and generate switch-like responses [66], robust oscillatory behavior [67], and dose-aligned responses [68]; networks that are capable of self-organizing cell polarization [69]; network motifs that buffer front-to-back signaling in polarized neutrophils [70]; and networks that respond optimally to time-periodic stimulation [71]. To these classic examples, we can add more recent contributions, such as integral feedback [72], feedforward loops [73,74], and insulators [31]. Using a complementary approach, many studies have focused on characterizing the functions that can be achieved by specific network topologies, with an important role of the parameter space (Figure 1.2); for example, the incoherent feedforward loop was shown to produce time-dependent biphasic responses and dose-dependent biphasic responses for mutually exclusive regions of the parameter space [75].

What Should Be Modeled? Future modeling work should include the examination of many questions, including but not limited to

1. What is the idealized dynamical behavior of a given module in the absence of interference from other network elements? In many cases, this is already known, but in order to build a fully dynamical theory, some points that are usually neglected should

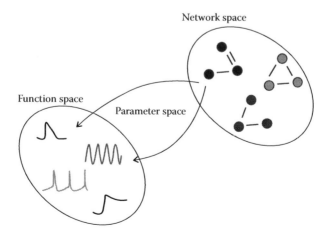

FIGURE 1.2 The same network topology can produce different outcomes in the function space depending on the set of parameter values.

be addressed. For example, how does a signal transduction cascade acting as an ultra-sensitive switch respond to an arbitrary time-varying input, or what are the delays associated with changing states in a bistable switch?

2. Under what conditions does a module maintain its idealized behavior when interacting with many inputs and outputs from other parts of the network? There are cases where the idealized behavior is quantitatively unchanged and cases where quantitative changes are observed, but may be accounted for by appropriate choices of effective parameter values. The conditions identified would include parameter regimes, classes of time-varying stimuli [30], and even choices of alternative network topologies. We expect that in some fraction of cases, we will find that the putative modules in fact almost never behave as true dynamical modules, and that in many more, the module will not be able to maintain its isolation for all choices of parameters and dynamical inputs. The latter cases will provide important clues as to how and when undesired cross talk can enter into signaling pathways, possibly resulting in pathological behavior. The former examples will suggest unexpected avenues for information flow even in wild-type networks. They will also lead the modeler to consider the following important question.

3. How might the introduction of insulators or changes in network topology create appropriately freestanding modules with the desired functions? To carry out these studies, a wide arsenal of techniques may be combined. These include singular perturbation theory to examine limits in which certain (combinations of) internal variables relax quickly [76,77], studies of system bifurcation diagrams as a function of both intended and retroactive inputs [78], methods of linear control theory to give quantitative local measures of retroactivity and cross talk [79], sensitivity analysis (both local and global), [22,80–85], ideas from disturbance rejection in nonlinear systems [76,86], and large-scale numerical screens, in which we will validate our analytical calculations by comparing simulations of a full network model with our predicted modular behavior for randomly chosen parameter sets.

1.2.5 Targeted Drug Design: Modeling Realistic New Applications to Defeat Cancer

Studies on modularity and retroactivity in cell signaling have the potential to uncover new mechanisms that lead to aberrant signaling in cancer, such as cross talk through retroactivity due to loss of insulation of some of the cycles in a signaling pathway [59]. We expect that the characterization of retroactivity and especially the discovery of fundamental principles of insulation will provide invaluable information for targeted drug design. Specifically, we pose that the local structure of the signaling pathway in which the target is involved carries crucial information about the efficacy of a drug directed against it. In fact, if the target of the drug is produced by a signaling motif that works as an insulator, it will be strongly regulated and thus insensitive to downstream retroactivity. As a consequence, the target may be difficult to modulate at drug doses that are clinically achievable.

The immediate translational value of the study of integrated signaling will be realized in novel insights on how to combine therapeutic approaches to best inhibit cellular behavior that is the result of aberrant integration across many pathways. The insights on modularity and information flow could be used to make predictions about the effects of pharmaceutical perturbations and to design combinations of interventions that effectively regulate inappropriate cell migration. An initial focus could be on off-target effects of clinically approved cancer drugs, with the goals of understanding deleterious side effects and predicting new therapeutic targets or new combinations for these agents; because the agents are already approved, such new uses could transition relatively quickly to the clinic. Subsequent work will determine to what extent specific time-courses of drug administration enhance therapeutic efficacy and identify combination therapies with additive or synergistic effects.

1.3 MODELING CANCER: THE GENOTYPE-TO-PHENOTYPE CONUNDRUM

The prevailing philosophy of the genomics age of cancer biology has been that specific genetic changes are responsible for driving cancer progression and tumor growth in specific ways that can be assayed and carefully teased out. Accordingly, the field has posited that understanding how specific genes contribute to these cellular processes will ultimately unravel the Gordian knot of tumorigenesis and reveal specific, druggable targets.

Unfortunately, the further we invest in the genome-wide analysis of cancer, the clearer it becomes that the cancer's genetic basis is far more complex than anticipated. As described above, (1) largely nonlinear cross talk between signaling pathways and (2) compensation by related signaling molecules increasingly suggest that cancer cannot be stopped by simply targeting single genes sequentially or even in pairs. While biologists continue working to develop new bioinformatic approaches to analyze the petabytes of sequencing data and myriad of signaling molecules and pathways that contribute to tumorigenesis and metastasis, mathematical modeling allows us the advantage of approaching tumor development from a phenotypic, rather than genotypic, platform. In doing so, we can attempt to reconcile the two approaches into a "unified theory of cancer," to borrow the metaphor from particle physics.

Modeling tumor growth on the basis of the phenotypic behavior of each individual cell within the tumor at least partially circumvents the issue of identifying how signaling pathways interact to drive cancer processes. When behaviors like invasion, motility, and proliferation are considered as end products, with inherent modulations in each property implicitly based on changing signaling inputs, scientists can probe functional questions such as "How fast will a tumor grow in a given microenvironment?" "Will a new drug actually select for more invasive cancer cells?" "How will a particular drug affect cell motility?" The results of such experiments in turn help inform biological hypotheses regarding the signaling molecules and pathways contributing to the emergent behaviors of these models.

Agent-based modeling (ABM) is a discrete-based approach, which can explicitly represent individual cells in space and time, and track and update their internal states according to a predefined set of biological and biophysical rules. The ABM has proven highly successful in identifying biologically relevant emergent behaviors, as well as explaining observed biological phenomena, in a variety of cancers (reviewed by [87–92]). ABM "simulates the (inter)actions of autonomous individuals within a complex system" [91]. Accordingly,

ABM has been applied to a broad range of disciplines (social sciences [93], ecology [94], and economics [95], to name a few), the unifying feature being the need to include heterogeneity among individuals to identify emergent system behaviors.

What Should Be Modeled? More detailed modeling of the dynamic, physiologic behaviors of individual cells should significantly improve tumor-scale models, as well as provide a more mechanistic link to gene expression changes in tumor cell populations. The majority of modeling approaches thus far focus on cellular responses to changing concentration gradients (of oxygen, nutrients, and other chemoresponsive factors) or whole cell responses (e.g., changing proliferation or apoptosis rates), but do not delve into the physical properties of cancer cells, such as cell shape changes, cytoskeletal remodeling, or adhesive forces between cells and their microenvironment.

The need for an enhanced, mathematically translatable understanding of the details of cell motility has been appreciated in the field [89] and is viewed as a requirement for accurately modeling rapidly migrating cells [96]. It is our hope that discoveries such as these can be applied to new and existing models of tumor growth, and that incorporating subcellular behaviors into tumor-scale models allows us to further bridge the gap between phenotype and genotype. A few groups have already started to link molecular signaling to tumor growth multicellular phenotypic behavior (the interested reader should refer to [97,98] for detail).

Possibly the greatest challenge facing the modeling field, and the most important determinant of future funding and expansion of the field, is that of efficient, functional communication between mathematicians and experimental biologists. In ideal situations, biological experiments feed descriptive, semiquantitative, or quantitative data into modeling endeavors, and the modeling results then reciprocally inform new biological hypotheses and experiments. In actuality, however, the needs and intentions of each community are often lost in translation. Compared with mathematics and engineering, biology has been a largely qualitative field, and accepted biological observations frequently do not easily integrate into mathematical models. Conversely, the modeling community can become bogged down in the interesting intellectual aspects of their research, while losing sight of the applicability to real biological systems and public health. This issue is not a new one, and the idea of integrated wet and dry labs has been proposed for quite some time [99].

1.4 CONCLUSIONS

Cross-disciplinary cancer research is inevitable in the next few decades. Reductionist approaches have revealed much useful information, but they have also taken a long time to demonstrate that they are severely limited. The days of expecting single genes to explain diseases like cancer are over. So what can we do to accelerate progress? First, we must teach from the earliest levels of biology training that cross-disciplinary work is essential. Second, we must create many fora for presentation and publication of modeling work that is integrated to biological reality, with the "Mathematical Oncology" section of the journal *Cancer Research* being one such auspicious forum, and we need others where hypotheses, theories, and different types of models can be carefully explained to the nonmathematical expert. Third, and the most controversial recommendation, we need more patient data—not more Western blots or gene arrays. We need more longitudinal patient data with outcomes and markers. We never studied cancer as we did other chronic diseases longitudinally, so we really do not have a major true survival index for the robustness of biomarkers in the human disease of cancer. It is time that we undertake these studies not in just highly resourced areas, but in low- and middle-income regions of the world where cancer incidence and deaths are rising, cancer is being diagnosed late, and risk factors and operative pathways are yet unknown, and where we can perhaps make a broad and expeditious impact in cancer mortality.

CHAPTER 2

Developing More Successful Cancer Treatments with Physical Oncology

With Joseph D. Butner

Complex biochemical and biophysical barriers inhibit the efficacy of chemotherapy, and the likelihood that chemotherapy will fail as a result of drug resistance in many forms is high. Clearly a more concerted, concentrated, and collaborative effort is required between oncologists, pathologists, chemists, applied mathematicians, and physicists if progress is to be made in the fight against cancer. Physical oncology has the potential to bridge the gap between these disciplines, quantifying through mathematical modeling the biophysical conditions affecting cancer treatment and tumor growth at multiple scales within the body, from the molecular (nano-) scale to the macroscale (whole body systems).

2.1 CHEMOTHERAPY

Current protocols for treating solid cancer tumors with chemotherapy or other treatment options, while not identical for all patients, commonly occur as follows: a patient is found to have a mass by his or her physician after either presenting with symptoms related to the mass or having an abnormal lab or imaging test that was unrelated to the mass; the patient undergoes further tests to determine whether a tumor is malignant or benign; if it is malignant, tests are done to determine how far the cancer has spread, and other tests may be done to determine its biological properties (mutations or protein expression). Once a baseline disease burden and diagnosis are established, the patient and physician discuss treatment options, including—but not entirely limited to—surgery, radiation therapy, chemotherapy, hormone therapy, and immunotherapy. Additionally, a combination of these therapies may be selected based on the patient's overall health, the extent to which the cancer has developed, the possibility of adverse side effects, and the patient's goals for treatment.

When chemotherapy is chosen as a treatment option, the decision is usually based on the goal of the patient and physician to prevent the cancer from progressing, to improve cancer-related symptoms, to prolong life, to reverse the spread of the disease prior to surgical treatment, or to cure the cancer. Generally, chemotherapy works by targeting the mechanisms that are necessary for cell division, such as the DNA or the cytoskeleton. This favors targeting fast-replicating cells (including cancer) and disrupting their growth. While the drugs affect cancer cells, they also affect the cells of the bone marrow, immune system, liver, hair follicles, gastrointestinal tract, and nervous system—almost any type of cell can be affected by the chemotherapy. The practice of delivering chemotherapy drugs to the patient involves intravenously issuing a drug or combination of drugs that are thought to be or have been proven effective in killing cancer cells on some experimental level. Some chemotherapies may also be injected directly near a tumor through a catheter or infused into the abdomen in specific situations. The results of the initial treatment are gathered using lab and imaging tests, and the empirical evidence is used to plan the next treatment.

Without being able to observe every single cancer cell in the patient, it is nearly impossible to determine with certainty if all the cancer has been killed. Cancer can return with the presence of only one surviving cancer cell, and indeed, incidences of cancer recurrence after chemotherapy are high. Due to this problem—the lack of certainty about how much of a tumor will be killed after a course of chemotherapy—we believe there is a strong need for mathematical modeling to predict the response of cancer to chemotherapy and other cancer drugs such that tumor treatment responses may be optimized relative to the response to current treatment standards.

2.2 DRUG RESISTANCE

Chemotherapy as an option for cancer treatment is limited by multiple drawbacks, namely, the negative side effects of cancer-fighting drugs due to the damage chemotherapy causes to healthy tissue and the cancer's ability to survive the drug therapies. Drug resistance is the cause of 90% of all chemotherapy failures, and is likely the cause of poor treatment outcomes, such as in breast cancer, where only 26% of patients with distant metastasis survive past five years [5,6]. Currently, clinical oncology is dominated by genetics and molecular biology. As such, the failure of chemotherapy is often attributed to mutations and abnormal proteins (see Figure 2.1 [100] for an example). Another obstacle to the success of chemotherapy is resistance to drug perfusion caused by physical barriers in the tumor, preventing the drug from penetrating across the abnormal tissue into cancer cells with adequate concentration required to kill the cancer. Essentially, if the drug cannot reach its target, it will not have an effect. That is where physical oncology comes into play.

Cancer therapy drugs must first enter the bloodstream, pass through the vessel walls, cross the cell membrane, and often even enter the nucleus of the cell in order to disrupt the DNA of the cell, either killing it or preventing it from dividing. Furthermore, each tumor resides within its own unique microenvironment, which consists not only of the tumor itself but also

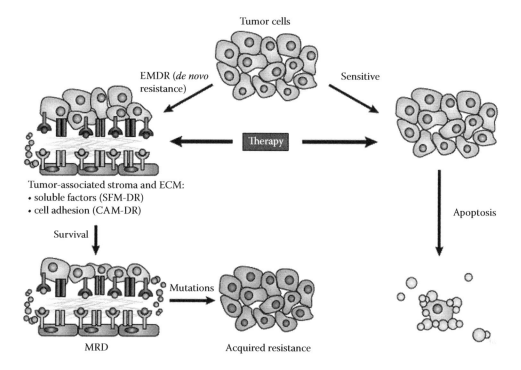

FIGURE 2.1 Illustration of one possible path for acquired drug resistance. A subset of tumor cells survives the therapy, resulting in minimal residual disease (MRD). Subsequently, genetic changes over time in these persistent cancer cells cause the gradual development of more complex, diverse, and permanent acquired-resistance phenotypes. These cells eventually cause disease recurrence and are much less likely to respond to subsequent therapy after acquired resistance develops. Environment-mediated drug resistance (EMDR); soluble factor-mediated drug resistance (SFM-DR); cell adhesion-mediated drug resistance (CAM-DR). (Reproduced with permission from Meads, M. B. et al., *Nat. Rev. Cancer*, 9(9), 665–674, 2009.)

the extracellular matrix (ECM), vasculature (or lack thereof), interstitium, and surrounding cells, each of which presents a potential barrier to transport of the drug to the cancer cell.

Thus, "physical drug resistance" represents a broad concept. It can be applied to hundreds of phenomena occurring within the body at multiple scales, from within individual tumor cells to whole tumor tissues and cancer that has spread or metastasized to other systems within the body. Reasons for resistance to cancer-fighting drugs can include genetic alteration within cells, compensatory biological pathway mechanisms, biochemical agents, altered effectors in DNA repair, pH alterations (acidosis), and removal of chemotherapy drug from the cell by cell membrane efflux pumps at the molecular level. This book focuses on physical barriers that exist at multiple scales within the body that impair the movement (or transport) of the drug from the blood, through the surrounding tissues, and into the tumor cell. Ultimately, a drug must reach its target (the DNA, microtubules, or other proteins) to have an effect. By understanding this transport process of how the drug reaches its target and what methods a cancer uses to impede this transport, we hope to improve the outcomes of cancer treatments.

2.3 FORMS OF DRUG RESISTANCE AND RELEVANT MODELING STUDIES

How then can physicians fight cancer with chemotherapy when there are so many conditions affecting the resistance of drugs that are proven to kill cancer in lab studies but are so often ineffective in live patients? Physical oncology attempts to provide physicians with patient-specific information that can lead to better-designed treatment approaches, with quantifiable predictions about the growth of tumors, the efficacy of drugs, and even survival rates of patients. Physical oncology takes into consideration not just the factors leading to drug resistance described above, but, more importantly, detailed physical descriptions about the microenvironment within an individual patient's tumor that may cause resistance to drug transport across physical barriers. *While most discussions about drug delivery are centered on what happens when the molecule enters the cancer cells, a very important (but less discussed) factor is the existence of multiple physical barriers preventing the drug molecules from even reaching the cell.* Limitations on the diffusion and transport of the drug can result in delivery of the drug in concentrations that are lower than the required lethal amount. All biological systems, including cancer, must obey physical laws such as mass conservation. As such, the delivery of drugs can be predicted, in part, by measuring the physical barriers of an individual tumor. By developing methods to do this in patients, we can begin to personalize cancer therapy and tailor treatment to optimize the individualized physical conditions for maximal results. To accomplish this, it is necessary to understand the various physical barriers to drug transport in tumors.

2.3.1 Genetic Alteration

Genetic alteration within cancers that have been exposed to the first round of chemotherapy is a commonly identified phenomenon leading to the proliferation of chemoresistant tumors [101–103]. Discouragingly for the patient, cancerous tumors that have responded favorably to the initial rounds of chemotherapy can return with an increased resistance if they are not killed entirely in the first round. The cause for this phenomenon is multifold, but one main reason is the proliferation of cancer cells within a tumor that have survived the first round because of genetic or phenotypic alterations. The offspring of drug-resistant survivors are resistant to subsequent rounds of chemotherapy. Insights into a specific patient's drug-resistant mutations within specific tumors can aid in the development of new mathematical models that predict kill rates based on those parameters, which can be helpful information for physicians to design chemotherapy treatments.

Foo and Michor [104,105] created a set of stochastic mathematical models describing the evolution of tumor cells, arising from a single genetic alteration, while being exposed to chemotherapy. Their models predict the probability that cancer will develop drug resistance during specific multidrug dosing regimens (Figure 2.2). The model output suggests candidates for the most efficacious dosing methods to treat metastatic tumors. This is just one example of how mathematical modeling can offer physicians optimum strategies to fight cancer that is inoperable or is otherwise best suited for chemotherapy. Other modeling work by Enderling et al. [106] used a two-part stepwise genetic mutation model to study how accumulation of harmful mutations may be involved in transition to a cancerous breast

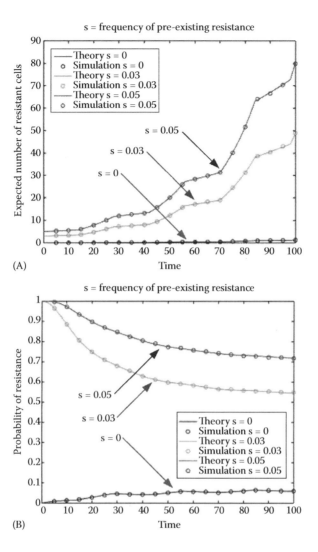

FIGURE 2.2 Preexisting drug resistance affects the dynamics of treatment response. Simulated (A) number of resistant cells and (B) probability of resistance are shown as a function of time for different preexisting resistance fractions under a certain therapy schedule. Simulation results are plotted as circles and theoretical formulas as solid lines. (Reproduced with permission from Foo, J., and Michor, F., *J. Theor. Biol.*, 263(2), 179–188, 2010.)

cancer phenotype. By comparing known genetic mutation rates in breast stem cells, they found that the prevalence of breast cancer is likely due to many possible mutation pathways, as the large number of disease instances is unlikely to occur due to only one mutation pathway. In this example, modeling has resulted in new insights on disease initiation that would be difficult, if not impossible, to obtain using traditional wet lab techniques.

2.3.2 Efflux Pumps

Another factor in chemotherapy resistance stems from cancerous cells' ability to expel drug molecules through the cell membrane with two forms of "pumps," consisting of proteins in the cellular membrane that carry molecules out of the cell before they enter the nucleus and the DNA is affected. P-glycoprotein, multidrug resistance-associated protein (MRP), and breast cancer resistance protein (BCRP) are found within the cell membrane and protect the cell's interior by expelling pathogens. These proteins function in healthy tissues to protect the cell from pathogens; however, in tumor cells, the overexpression of these proteins can protect the cells against drugs and lead to drug resistance. These proteins are currently the targets of several pharmacological approaches to improve the efficacy of chemotherapy by allowing the drug molecules to remain within the cell through deactivation of inhibition of these proteins. Atari et al. [107] developed a mathematical model to predict resistance to the chemotherapy drug topotecan in breast cancers containing high concentrations of BCRP. This model is an example of how awareness of certain parameters—in this case, the specific concentration of BCRP—and their application to mathematical models can aid in predicting the optimum doses of a drug to eradicate cancers in individual tumors in specific patients. Use of the model in future design of optimal dosing strategies may prevent recurrence of cancerous cells due to resistance to the drug after the first round of treatment.

2.3.3 Cell Cycles

Cell cycles also play a role in the cell's ability to resist chemotherapy. Mathematical approaches designed by Roe-Dale et al. [108] model cell cycles and quantitatively explain the effect of cell cycles on resistance to multiple drugs administered in different sequences during treatment, namely, doxorubicin (DOX) and cyclophosphamid, methotrexate, and 5-fluoruracil (CMF). Chemotherapy drugs can disrupt cancer cells at different points in their growth cycle, but certain drugs are more effective in impeding cell proliferation at specific points in the cycle. If a specific drug is absorbed into the cell at a point in its cycle when it is less affected by that drug, then there is a greater likelihood that the cell will survive and proliferate. This resistance to chemotherapy can be overcome if patients are given sequential doses of alternating drugs (i.e., DOX and CMF), which disrupt cell proliferation at different points in the cycle (see Figure 2.3 for an example). This two-compartment model describes the interaction of cell cycles, drug treatment, and drug resistance, and created simulations consistent with previously observed 10-year clinical outcomes of patients described by Bonadonna et al. [109]. The model was effective in predicting drug resistance and treatment outcome for future patients and clarified differing results in patients who were administered the drugs in different sequences. Drug resistance is a key determinant in the success of chemotherapy, but insights into the relationship between drug resistance and cell cycle stages gained from mathematical modeling can help physicians develop treatments that are helpful in improving patient outcome.

2.3.4 Acidosis

High acidity levels often exist in tumor microenvironments where cancer cells have produced an acidic environment through anaerobic glycolytic metabolism due to a lack of oxygen reaching these cells. Tumors can flourish in this environment, which exacerbates drug resistance due to chronic inflammation, inhibits specific immune responses, and can

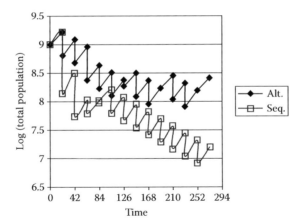

FIGURE 2.3 Change of surviving tumor cell population over time under treatment. Diamond, alternating regimen; square, sequential regimen. The sequential regimen keeps the total population below that for the alternating case in most of the time points. (Reproduced with permission from Roe-Dale, R. et al., *Bull. Math. Biol.*, 73(3), 585–608, 2011.)

trigger angiogenic growth factors (and many other growth factors) that help tumors grow, survive, and even metastasize [110]. Patients with high acidity levels in the microenvironments of a variety of cancers have been observed to be less likely to survive, as the acidic environment allows the cancer cells to proliferate and induces many changes at the genetic and molecular levels that are very beneficial for the tumor. Drugs formulated to carry a certain charge in relation to the physiological pH may not interact as intended with cell membranes in extremely acidic environments, as mass transfer across cell membranes is dependent on favorable pH conditions. Lee et al. [111] have applied mathematical modeling to the study of cancer motility, or movement from the point of origin to areas of tissue with more favorable conditions, with consideration of environmental factors influencing proliferation, such as acidosis and hypoxia (i.e., depravation of oxygen) created by the increasingly complex vasculature, anaerobic glycolytic metabolism, and interstitial pressure of the tumor microenvironment. Acidosis therefore creates an environment of increased drug resistance, as well as conditions leading to the spread of cancer to new areas.

2.3.5 Hypoxia

Silva and Gatenby [112] created a mathematical model to understand the evolutionary dynamics influencing drug resistance in a tumor microenvironment. Their model showed that the hypoxic environment at the center of a tumor mass results in increased drug resistance, where increasing doses may have no beneficial effect on the patient, but may increase patient toxicity. Interestingly, the model also confirmed earlier models that suggest that "containing" the cancer might be more beneficial to patient survival than the goal of eradication. Frieboes et al. [113] implemented a mathematical model of the response of a tumor to chemotherapy taking into account the internal mass concentrations of three parameters: the drug, oxygen, and nutrients. By combining biological data with computational modeling, the model was able to predict resistance to the drug, and was helpful in developing more advanced methods for improving future treatment outcomes. While many factors influence resistance to chemotherapy, the authors mainly seek to use computational modeling to explain how conditions—physical and biological—within the tumor create barriers to the transport of chemotherapy into tumor cells.

Acidosis is commonly found together with hypoxia, as the lack of oxygen commonly results in a transition to anaerobic glycolytic metabolism. Anaerobic metabolism commonly occurs in healthy systems during periods of strenuous exertion, after consumption of all the readily available oxygen, resulting in lactic acid buildup in the muscles, which commonly leads to soreness in the muscles later. Similarly, anaerobic glycolytic metabolism occurs in regions of a tumor where oxygen supply is limited, leading to both conditions concurrently in the same location. Mathematical modeling by Gatenby et al. [114] and Smallbone et al. [115] has provided valuable insights in the development of these conditions, where it was observed that increased distance from oxygen supply resulted in a highly acidic microenvironment. This leads to a selective pressure, where cells with an acid-resistant phenotype and a propensity for anaerobic glycolytic metabolism were more likely to survive. This hardier phenotype has a selective advantage over both other cancerous phenotypes and healthy cells, and may damage surrounding healthy cells as they continue to produce an increasingly acidic microenvironment.

2.3.6 Interstitial Fluid Pressure and Electrostatic Charge

Between the various functioning portions of an organ is a space called the interstitium. Each organ has its own unique interstitial space consisting primarily of stromal cells, which are connective tissue made of cells such as fibroblasts (involved with cellular metabolism and cell repair) and pericytes; inflammatory cells; and molecules that are involved in cellular biochemical and structural support, known as the ECM. Within a tumor, it is common to find an environment hostile to normal cells due to hypoxia, acidosis, low glucose levels, and low levels of the energy carrier adenosine triphosphate (ATP), in addition to a high density of the ECM described above. Leaky vasculature contributes to increased pressure in this environment by allowing fluid into the interstitium. Figure 2.4 [116] shows the distribution of interstitial fluid pressure (IFP) levels in tumors.

Mathematical modeling in collaboration with *in situ* analysis has been successfully implemented to describe the relationship between interstitial pressure and its lymphatic drainage system, which relieves increased interstitial pressure [117]. In a tumor environment, the lymphatic system is also impaired, and the higher intratumor pressure combined with the dense tumor environment impairs blood flow, creating an environment through which it is difficult for drugs to penetrate. Specific measurements of this environment within a live patient's tumor are required in order to predict how well chemotherapy can diffuse into the tissue. When the information is available, it becomes instrumental in calculating accurate concentration, volume, and delivery time of the drug. Stylianopoulos et al. [118,119] created mathematical models of the tumor interstitium to predict the outcome of the interaction between charged drug molecules and the electrostatically charged,

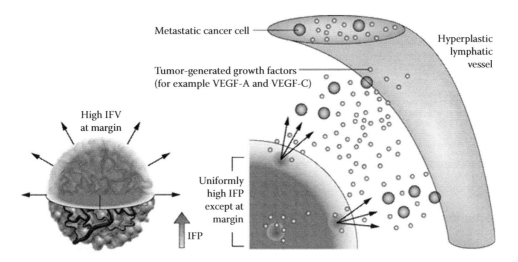

FIGURE 2.4 Elevated IFP levels in tumors. The IFP is uniformly elevated in tumors, but at the margin, there is a steep drop of IFP. This steep drop causes body fluid, tumor-generated growth factors, and cancer cells to leak out of the tumor into the peritumoral tissue, which may in turn facilitate angiogenesis and metastasis, and inhibit drug delivery. Interstitial fluid velocity (IFV); vascular endothelial growth factor (VEGF). (Reproduced with permission from Jain, R. K., and Stylianopoulos, T., *Nat. Rev. Clin. Oncol.*, 7(11), 653–664, 2010.)

densely packed cellular matrix in tumors with closely aligned fiber networks. Their model predictions suggested that altering the electrostatic charge of delivered molecules and nanoparticles (NPs) can affect the diffusivity of particles in tumor tissue. By analyzing fiber network orientation with different degrees of fiber alignment, their results demonstrated that diffusion anisotropy becomes even more significant with increasing degree of fiber alignment, particle size, and fiber volume fraction. Figure 2.5 shows how the diffusivity of NPs changes with fiber volume fraction.

2.3.7 Angiogenesis and Vasculature

Most tumors are also characterized by a high level of microvasculature with abnormally high IFP and heterogeneous blood flow [120]. Tumors experience extensive angiogenesis as they strive to supply more and more blood to an environment hostile to circulation. Elevated IFP, paired with increased microvascular pressure, may lead to physical collapse of the blood and lymph microvessels, resulting in poor fluid drainage, poor transport of oxygen and nutrients, and poor drug delivery. Coupled with these biophysical barriers are the biochemical gradients described above. Figure 2.6 shows angiogenesis in a tumor obtained using an advanced form of optical imaging.

A series of hybrid models has been developed by Lowengrub's group to study tumor formation, vascular modeling, and angiogenesis [122–125]. Wu et al. created a vascular tumor growth model that combined a continuous growth model (representing earlier work on modeling tumor growth) with a discrete angiogenesis model to display the effect of interstitial pressure on vascular tumor growth [124]. The model shows how blood flow

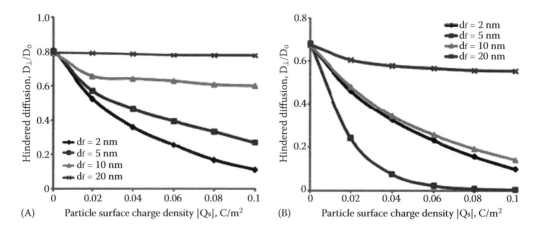

FIGURE 2.5 Effect of fiber volume fraction on the diffusion of NPs. Two fiber volume fractions: (A) 0.03 and (B) 0.06. As the fiber volume fraction increases, the fibers are packed closer to each other. Hindered diffusion is the ratio of the overall diffusion coefficient in the fibrous medium transverse to the fiber direction over the diffusion coefficient in solution. In (A), the diffusivity increases as the fiber diameter increases; however, in (B) the diffusivity for 5 nm fibers is lower than that for 2 nm fibers. d_f, fiber diameter. (Reproduced with permission from Stylianopoulos, T. et al., *Biophys. J.*, 99(5), 1342–1349, 2010.)

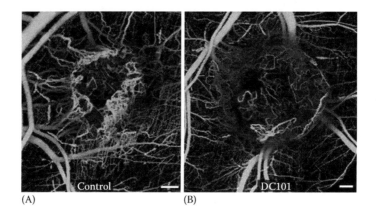

FIGURE 2.6 Visualization of tumor microstructure and vascular morphology. Images for control (A) and treated (B) tumors. (Reproduced with permission from Vakoc, B. J. et al., *Nat. Med.*, 15(10), 1219–1223, 2009.)

within the tumor—and therefore delivery of oxygen, nutrients, and drug molecules—is limited by both internal pressure and blood vessel collapse. The model was then extended to understand if the high IFP would also reduce therapeutic drug efflux into the tumor (Figure 2.7) [125]. It was found that as the tumor shrinks, the IFP profile becomes nonuniform, slowing down the treatment response at later stages due to the low drug concentration in the tumor interior. This model also holds implications for how physicians' treatment strategies could take into account the increased resistance to physical mass

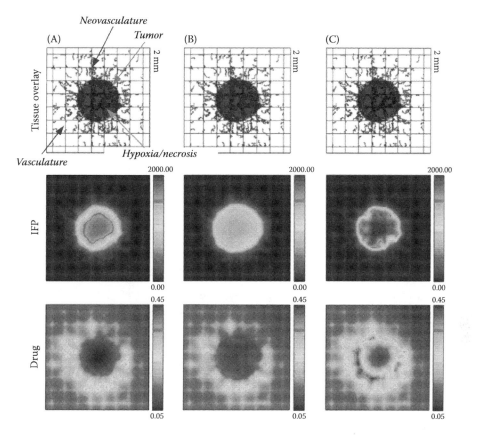

FIGURE 2.7 Examination of heterogeneous tumor dynamics due to chemotherapy drug delivery. A rescaled, simulated treatment at approximately day 18 is shown. (A) Control. (B) IFP with elevated tumor interstitial hydraulic conductivity. (C) IFP with elevated tumor vascular hydraulic conductivity. Drug concentration levels in and near the tumor: (C) > (A) > (B), where the drug distribution is more heterogeneous in (C) than in the control (A). Tissue overlay display: Tumor (viable, hypoxic, and necrotic tissue) and the preexisting vasculature (rectangular gridlines) and neovasculature (irregular lines). IFP display: Scale bar, pressure in pascals. Drug display: Nondimensional unit. (Reproduced with permission from Wu, M. et al., *J. Theor. Biol.*, 355, 194–207, 2014.)

transport due to elevated IFP. Together, these models provided insight into questions about how much blood is capable of flowing into the tumor, how the pressure within the tumor affects drug delivery, and how these conditions may lead to further tumor growth and metastasis. Mathematical modeling by Jain et al. [126] has also been used to gain insight into the effects of vasculature normalization (a process where antiangiogenics treatment is used to temporarily "normalize" vasculature in tumors back to a state closer to that in healthy tissue) on drug delivery into the tumor microenvironment. Their results suggest that antiangiogenic therapy may help increase drug delivery in tumors due to increased convection within the tumor interstitium and decreased bulk loss of delivered drug from the tumor environment.

2.4 PATIENT-SPECIFIC PHYSICAL PROPERTIES

2.4.1 Mass Transport in Tumors

By creating mathematical models of these phenomena across biological scales (e.g., blood vessels, ECM, and cellular membrane transporters), we have been successful in offering *predictive* analysis of the individual tumor's response to drugs, as well as tumor growth or death. For example, we have used principles of physics and conservation laws to predict how much drug is required in a specific patient to kill a certain volume of cancerous cells within a specific tumor. Along these lines, we published a paper entitled "Mechanistic Patient-Specific Predictive Correlation of Tumor Drug Response with Microenvironment and Perfusion Measurements" in 2013 in the *Proceedings of the National Academy of Sciences* [7]. This work drew attention for its accuracy in predicting the penetration of drugs into liver metastases from colorectal cancer.

What this study suggested is that it should be possible to mathematically predict how much cancer will be killed through an individual patient's prescribed chemotherapy regimen, by measuring the physical properties of the tumors that relate to blood perfusion and diffusion distances from the blood vessels to the cancer cells. These predictions are mainly based on patient-specific parameters derived from contrast-enhanced computed tomography (CT) scans and histopathology—parameters that are used to inform exact mathematical solutions of conservation equations that we developed and applied to clinical studies. The corresponding clinical outcomes compared with the predictions provided by the model help us understand drug resistance. Without an idea of the resistance to drug transport within the specific patient's tumor and its microenvironment, it is difficult to successfully kill the cancer according to either traditional or novel and experimental chemotherapeutic practices.

Applying these new quantitative tools can help reveal the physical mechanisms that cause resistance to chemotherapy and inhibit the efficacy of drug therapy. We hope that these novel approaches to understanding all the factors of the body's resistance to chemotherapy—in both the research being conducted around the country and that which is being performed by our groups—will improve treatment approaches and change the current outlook for cancer patients who are otherwise faced with discouraging prospects. The new studies driven by the success of our models are the subject of Chapter 5.

2.4.2 Diffusion in Unique Tumor Microenvironments

Diffusion is simply the random movement of a substance—in this case, a particular amount of drug molecules in the interstitium—from a region of high concentration to a region of low concentration. The area across which the drug must move is characterized by a mathematical coefficient that describes the ease with which the substance can move dependent on the density of the medium and any other physical barriers present.

The ECM within most tumors is very dense. Human tissue in all parts of the body is not simply made of individual cells that perform the function of the particular organ, but is composed of the cells plus a network of molecules that connect and support the cells, called the ECM. The matrix in each organ is unique, and the function of the ECM is to support the tissue, guide growth and development of the cell, and regulate the intercellular communication. *In cancer, the ECM surrounding the cells can be quite dense; therefore, it creates an environment in which cancer-fighting drug molecules have difficulty reaching their intended target in the cancer cells.* Furthermore, the tumor microenvironment is unique to each tumor in each patient. As tumors develop from a single mutated cell, the mass grows among the normal cells, characterized by an abundance of rapidly dividing cells in a small area. As the tumor cells continue to grow abnormally and replicate, they may require more blood circulation, which leads to angiogenesis—the growth of new blood vessels. Each tumor in every individual patient exists at a different stage in its development when it is diagnosed; therefore, developing strategies to treat cancer in a patient requires quantifiable data about the specific tumor targeted for treatment.

In our recent work, we have shown how individual patient data about specific tumors derived from contrast-enhanced CT scans and histopathology have been used to validate equations written to predict cell kill within the tumor. By deriving specific measurements of a patient's tumor, it is possible to use the data in designing a treatment strategy that will attack specific tumors while minimizing toxicity to the surrounding cells. Because each tumor will have its own specific distribution of size, ECM density, IFP, and surrounding vasculature, each tumor will behave differently when exposed to drugs that are meant to perfuse and diffuse through the microenvironment. Understanding this element of drug resistance is instrumental in the design of successful patient-specific approaches.

The use of biomarkers for individual patients gained from contrast-enhanced CT scans in combination with mathematical models will be helpful in understanding how tumors respond to new chemotherapy drugs or immune therapies. This integrated approach may also help in early detection of cancer metastasis. After immune therapy, scans may be repeated to determine the progression of cancer or response to treatment, and measurements done on tumors that determine the diffusion gradients based on staining levels can assist in diagnosis or informing for further treatment. Measurement of biomarkers within the tumor will also allow mathematical prediction of responses to any cytotoxic agents used in drug therapy. This emerging science will hopefully lead to a more targeted and individualized approach to cancer treatment.

2.4.3 Modeling at Multiple Scales

There are many reasons to account for multiscale effects when investigating cancer and developing successful cancer prevention and treatment strategies. Cancer growth is an emergent, integrated phenomenon that spans multiple spatial and temporal biological scales resulting from dynamic intracellular interactions and interactions between individual cells, and between cells and their constantly changing microenvironment (Figure 2.8) [97,98,127–130]. All these statements, which are all true, implicate that cancer growth and treatment are all multiscale processes that act at the molecular, cellular, tissue, and whole body scales [131–137].

Other than these common statements, to us, another compelling reason is that, from a physics perspective, the timescales for drug delivery and tumor response to treatment

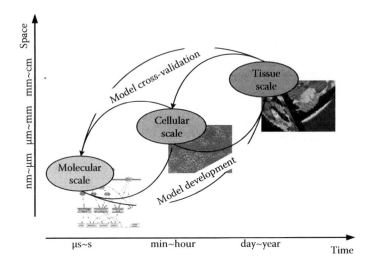

FIGURE 2.8 Schematic illustration of a cancer model that spans three biological scales: molecular, cellular or multicellular, and tissue scales. Different scales represent different spatial and temporal ranges, and each of these spatial scales may have multiple temporal scales. For example, molecular scale (nanometers to micrometers, microseconds to seconds) can refer to protein-based signaling pathways, cellular or multicellular (diffusion) scale (micrometers to millimeters, minutes to hours) can refer to both single and multicellular properties due to cell–cell and cell–matrix interactions, and tissue scale (millimeters to centimeters, days to months) can refer to gross tumor behavior, including volume and morphology. Multiscale modeling requires the development of a functional linkage between these scales. Note that biological processes that take place at a smaller scale generally happen much faster than those at a larger scale.

processes (cancer cells may die or still grow or move under treatment) overlap—that is, there is no scale separation with respect to time (Figure 2.9). Thus, it is necessary to use a multiscale or multimechanism approach to "simulate" all these drug treatment–related processes together in one place in order to determine the correlations between our imaging markers and treatment outcome.

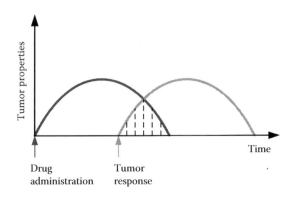

FIGURE 2.9 Simple schematic for explaining the overlapping of processes involved in drug treatment.

2.5 BRIEF REVIEW OF MATHEMATICAL MODELING OF CANCER TREATMENT

As aforementioned, cancer treatment is a complex phenomenon and involves many biophysical factors. Some of these factors have been successfully built into mathematical models, including chemotherapeutic drug dosages and dosing schedules [138–140], drug delivery and distribution within the tumor and tumor vasculature [141,142], tumor response to drug treatment [143], and adaptive therapy based on tumor response to previous treatments [144], as well as other treatment approaches, such as antiangiogenesis [145] and radiation therapy [146]. Others examine the effect of treating different types of cells on phenotype population dynamics, such as cancer stem cells versus differentiated cells [146–148] (see Figure 2.10 for an example). In each of these studies, simulation results have been compared to *in vivo* experimental data to confirm the model results that agree with outcomes observed in living systems. These promising results have been achieved using a wide variety of modeling approaches, including compartmental [146], continuum [143], pharmacokinetic [141], and hybrid [142] models, and over a wide range of cancer types.

In recent years, exciting new technologies have emerged to combat the shortcomings of traditional cancer therapies. Of these, one of the more promising is NP drug delivery, where porous or hollow NPs can be loaded with cancer-fighting agents and used to preferentially deliver their payload to the tumor [149] while reducing harmful exposure of healthy tissue to the cancer-fighting agents. While promising, NP delivery presents unique challenges above traditional chemotherapy delivery, where drug is injected into the blood and distributed along the entire circulatory system. Because NPs serve to compartmentalize the treatment agent, achieving preferential delivery to the tumor requires preferential accumulation of NPs within the tumor. Accomplishment of this goal requires detailed

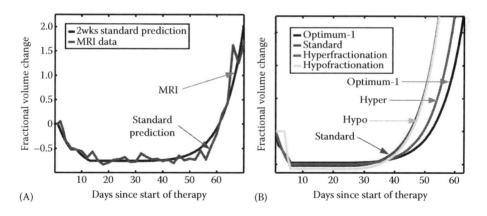

FIGURE 2.10 Parameterization and a time-dependent model for optimizing radiation dosing schedules. (A) Volumetric time-series MRI data of mouse gliomas after treatment with 2 weeks of standard therapy were used to quantify model parameters. Model predictions are in agreement with observed data. (B) Model predictions using the parameter values identified in (A) for all four treatment schedules: optimum-1, hyperfractionated, hypofractionated, and standard schedules. (Reproduced with permission from Leder, K. et al., *Cell*, 156(3), 603–616, 2014.)

understanding of how NPs move in the bloodstream and what causes them to preferentially aggregate. To this end, modelers have endeavored to describe this process mathematically, and have shown success in replicating results observed *in vivo* [11,150–154]. For example, interesting recent *in vivo* validated modeling results have successfully predicted liposomal distribution in microvasculature and across the tumor interstitium and intratumor liposome accumulation [150], provided valuable information about how the size distribution of NPs affects accumulation in the vasculature [151], yielded interesting insights about the need for NPs with longer drug release times [152], and helped demonstrate that targeted NPs allow for large concentrations of drug to reach the target, decreasing toxicity to the host in nontarget locations and decreasing the dosage [11]. In these cases, *in vivo* model validation is easily accomplished through the use of fluorescent NPs, which may even be viewed in real time in living systems using techniques such as intravital microscopy. Mathematical modeling has also indicated that positively charged particles have better drug delivery in tumor cells *in vitro* [155].

2.6 BRIEF REVIEW OF SOME PHYSICAL SCIENCES–ONCOLOGY CENTERS

How physical sciences are involved in the understanding of and battle against cancer is the focus of at least a dozen dedicated Physical Sciences–Oncology Centers (PS–OCs) located around the United States. Work being done in these research facilities pursues an understanding of how cancer functions according to physical laws and principles at multiple scales within the body, from the subcellular to whole body systems.

Researchers at the H. Lee Moffitt Cancer Center in Tampa, Florida, are seeking to apply mathematical modeling to three different projects: understanding how the physical microenvironment influences the formation of cancerous cells, how the microenvironment influences the ability of cancer cells to invade and grow in otherwise healthy tissue, and how models of the tumor microenvironment can be scaled up to the tissue scale [144,156]. Each of these projects employs actual specific individual patient data in predicting therapeutic outcomes for individual patients. The success of these modeling programs can be described as the ability to plug patient-specific values as parameters into the equations, with the potential clinical use being output that assists in individualized treatment planning.

Researchers at the University of Southern California (USC) PS–OC are focused on understanding two interrelated aspects of cancer: tumor growth and drug response. Modeling at the cellular scale is used to predict how cells will react to changes in their genetic makeup and in their local environment in response to therapeutic intervention. Dr. Cristini was coleader with Dr. Sam Gambhir for a project entitled "Multi-Scale Cancer Modeling: From Cell Phenotypes to Cancer Spread and Response to Therapy," which aims to provide insight into how lymphoma tumors grow and respond to treatment. A number of mathematical models have been developed to predict the lymphoma tumor's response to the drug by taking into account mass transport properties of the tumor, as well as the physical mechanisms influencing tumor growth [157,158].

At the Methodist Hospital Research Institute PS–OC, Dr. Cristini, along with Dr. Paolo Decuzzi, led a core project to simulate the behavior of small molecules and particles at the nanoscale within the microvasculature of the tumor environment. The goal of the project was to model and predict transport of chemotherapeutic particles into and through tumor microenvironments to understand how drug molecules will affect individual patients when physical measurements gained from patient imaging are applied to modeling equations as parameters [153,154].

In conjunction with these efforts, we have contributed to the understanding of physical oncology through mathematical modeling of patients with colorectal and pancreatic cancers. Our theoretical work and clinical studies have shed light on how chemotherapy behaves within tumors by applying principles of mass transport; these efforts are geared toward offering patients more individualized treatment.

Our work complements and draws from studies by other scientists around the world that are attempting to identify elementary biophysical laws of cancer development and develop new mathematical models of cancer. This work has drawn the attention of clinical and basic scientists alike, as our results are helping direct the future of physical oncology and modeling of cancer. Work in this area examines the entire process of treating cancer through chemotherapy (and other treatment options), from injection of the free drug into the bloodstream to the diffusion of drug through blood vessel walls into the tumor microenvironment. Quantification of barriers to diffusion, predictions of tumor growth, modeling of the physical properties of the complex tumor microenvironment, understanding tumor morphology, and multiscale modeling of tumor growth are key current topics within the field of physical oncology [159]. Each branch provides a quantitative tool that assists in furthering the overall understand of how cancer develops, how it resists drug intervention, and how it can be successfully killed. Modeling has also described the interactions between immune and other cells, and the migration of cells from areas of malignancy to other regions of an organ or metastasis to other body systems.

2.7 CONCLUSIONS

The complex biochemical and biophysical barriers inhibit the efficacy of chemotherapy, and it is highly likely that chemotherapy will fail as a result of drug resistance in many forms. Mathematical modeling of cancer and physical oncology in general is ultimately a pursuit of improved patient health and quality of life through more accurate diagnosis and targeted, patient-specific treatment. By gaining more detailed physical descriptions of individual tumors in patients, and by predicting the outcome of treatment using mathematical modeling, physicians will be able to better inform their patients about their prospects and will be better equipped to help with decision making, treatment plans, and prognosis. The models developed and tested by our research groups on a variety of cancer types have been accurate up to 85% in predicting cell kill rates, tumor growth, and even patient survival [7]. Imagine the change in the patient's state of mind when embarking on a chemotherapy routine if the patient and physician were able to predict the outcome of the treatment with the prospect of improving treatment success rates while reducing the exposure to more drugs than are necessary, thus avoiding side effects produced by drug overexposure.

Overall, the results of recent work done across the country to mathematically model cancer and cancer treatment outcomes offer encouraging prospects for improving patient outcomes. In order to accomplish the goal of improving quality of life, more dialogue is needed between the sciences of oncology, pathology, chemistry, physics, and engineering. Improving patient quality of life is the end result. However, other compelling and potentially desirable outcomes include significantly lower treatment costs and more efficient treatment, better understanding between physicians and patients to assist in decision making, and more efficient interaction between patients and doctors, who will have specific predictions and treatment plans designed to treat specific tumors existing within unique conditions with their own resistance profiles. Before we can arrive at this optimistic conclusion, much work must be done to approve and implement clinical studies that will help inform practitioners of physical science in oncology of the best mechanistic practices for cancer treatment.

CHAPTER 3

Mathematical Pathology

$\mathbf{I}$F PATHOLOGY IS "THE study of the essential nature of diseases and especially the structural and functional changes produced by them" [160], then "mathematical pathology" is the branch of this field in which the first principles of physics are applied to the study of diseases, including cancer. The objective of mathematical pathology is to mathematically model and explain the structural and functional *mechanistic* changes happening within the body as a result of the onset of disease. The foundations or "first principles" of physics, for instance, Newton's mechanical laws of motion, assign the same fundamental conservation laws to all matter, including the biological systems within the human body, for example, conservation of mass and of molecules diffusing through a tumor, which may often be treated mathematically as a closed system. Functional changes in the body brought on by the evolution of cancer can be explained mathematically, as described in this chapter. We first identify the justification for mathematical pathology as a necessary and useful step in the diagnosis of cancer, using the example brought by our recent collaborative work with Dr. Mary Edgerton (MD Anderson Cancer Center) on ductal carcinoma in situ (DCIS), the most common form of noninvasive breast cancer, which often precedes invasive ductal carcinoma (IDC).

3.1 BIOLOGY AND TREATMENT OF DUCTAL CARCINOMA IN SITU

DCIS is an early form of breast cancer in which the tumor originates within and is confined to the breast duct. The breast is composed of mainly adipose tissue and 10–20 main ducts that branch into smaller ducts whose primary purpose is to discharge milk. Growth initiates with a single mutated cell at the base of the duct [106] and progresses toward the opening of the duct, where it essentially stalls in its growth before breaching out of the duct—if it is to be considered DCIS. Should the growth extend beyond the duct, then the cancer most likely progresses to IDC [161], which accounts for 80% of all breast cancer diagnoses. While DCIS is not a fatal disease, it often progresses to IDC, which can be fatal. To combat DCIS, women may elect to have the entire breast removed or have breast conservation surgery, in which the portion on the breast that is identified by mammography to be affected by DCIS is removed; however, 20%–50% of breast conservation cases require an additional surgery to remove all DCIS, because the "surgical volume" is poorly defined by mammography [162].

When pathologists examine patient biopsies of breast tissue with DCIS, they are looking to see whether the neoplastic cells have stayed within the duct (*in situ* disease) or invaded through it. They are also looking at how aggressive the cells look (i.e., the tumor grade), among other features. DCIS has been described as having a short initial stage of fast growth in its development, followed by a longer stage of slow growth—or zero growth. By the time DCIS shows up on a mammogram, most patients will have had the disease for at least five months, as this is the time necessary for buildup of microcalcifications in individual ducts, which are then detectable by mammography. These microcalcifications appear as fine lines in mammograms, as if small thread fragments were scattered through the tissue. These are actually the remnant of necrotic cells found at the core of breast ducts and reveal the development of the cancer. Before planning surgery, a team of physicians (radiologists, pathologists, and surgeons) will determine the boundary of the breast tissue affected by DCIS and decide how much tissue needs to be removed. Oftentimes, immediately before surgery, a radiologist places a needle or other marker to localize the center of the tumor with the help of x-rays. In the operating room, the surgeon will then measure the amount of tissue to remove around the needle-localized tumor.

Determining the extent of disease is critical for a surgeon to decide how much margin of normal tissue to remove around the tumor. Younger tumors will have a viable rim characterized by cells in a state of proliferation. More advanced DCIS will be characterized by a rim with more cells in a state of apoptosis and a necrotic core of dead cell material that has built up along the central ductal axis as a result of the static pressure delivered by the proliferating cells at the rim of the tumor. Despite the effort to screen women for breast cancer once a year through regular mammography, current clinical practices in breast cancer detection and treatment have proven to be inadequate in preventing the disease from affecting patients [163], in part because the timescale for disease progression is shorter than current screening frequency.

3.2 NEED FOR MATHEMATICAL PATHOLOGY

Early detection of DCIS and breast cancer by mammography may result in better control and treatment of the tumor [164]. However, because the disease has usually progressed past the initial fast-growth phase by the time of detection, patients are often faced with the immediate decision of whether to proceed with breast conservation surgery or more radical surgery to remove the entire breast (in a procedure called mastectomy). In some cases, physicians need to discuss other treatments with the patients, such as radiation, hormone therapy, or chemotherapy. Recent studies have demonstrated that breast conservation therapy, in combination with radiation therapy, has disease treatment success rates comparable to those of full mastectomy treatments [165]. Because many women ultimately opt for breast conservation surgery when a full mastectomy is not clinically necessary, determining the optimal surgical volume for resection is essential for the patient's survival. Remove too much tissue, and the patient experiences the discomfort, disfigurement, and trauma of unnecessary surgery; remove too little of the affected breast, and the patient may need to undergo one or more additional surgeries to completely remove the affected area of malignant cells [166–168]. Methods to predict the true extent of the disease would help surgeons and patients decide the optimal surgical approach to treatment.

Hence, the goal of mathematical pathology in relation to DCIS is to accurately describe how DCIS develops to aid in therapeutic planning to define the volume of breast tissue that must be surgically removed, that is, the surgical volume. An accurate assessment of surgical volume is instrumental in providing physicians and patients all necessary information before surgery. We proposed the inclusion of a mathematical modeling-based mechanistic step in cancer treatment to be implemented in parallel with currently available prevention and detection practices to give patients and physicians a more complete understanding of the specific tumor they are fighting. Through mathematical modeling of cancer, it is possible to describe where a tumor is in its development, and, for DCIS, where the edge of the tumor is located, which aids in surgical planning. We designed our approach so that all data being supplied to the mathematical models could be derived from a patient's biopsy and integrated into the clinical workflow.

3.3 MATHEMATICAL MODELING OF DCIS

In order to describe the nature and development of specific tumors within a patient, specific values are required for parameters in mathematical equations. In our modeling work on DCIS, three key parameters are required, and they can all be assessed in pathological analysis of patient biopsies performed *at a single point in time* as a step in between detection through mammography and surgical planning. The three key parameters are related to tumor development; involve cell proliferation and death, as well as the physical ability of cells to receive nutrients depending on the diffusion properties of the microenvironment; and will provide output useful in determining the tumor margin. They are the *proliferative index*, the *apoptotic index*, and the nutrient *diffusion penetration length*. Before exploring the specifics of each study pertaining to this chapter, it is first helpful to define two terms and briefly explain their relevance to the efficacy of chemotherapy, as this is the subject of Chapter 5 on a novel mathematical approach to determining the fraction of tumor cells killed by chemotherapy.

Perfusion is, broadly speaking, the force-driven movement of a fluid through a solid substance. In medical cases, perfusion is the physical transport of a fluid such as blood through a vessel, for example, the heart or the vasculature of a tumor. In cancer modeling, perfusion is an important process in elucidating many phenomena, such as the transport of chemotherapy to a tumor. Perfusion is also relevant to cancer treatment in that each tumor microenvironment exhibits its own physical characteristics influencing perfusion dynamics, including the mechanics of delivery of molecules of interest (such as chemotherapy or oxygen) through the bloodstream to the tumor [169].

Diffusion pertains mainly to the movement of a substance from an area of high concentration to an area of low concentration; in medical cases, this occurs after the substance has been delivered to the tissue of interest through perfusion. When introducing chemotherapy through the bloodstream to target a tumor, the drugs must diffuse from areas of high concentration (i.e., the bloodstream or tissue near the vasculature), often *across the dense barriers of the tumor* into the entire mass of the tumor area [170]. In our physics classes, we all have seen that it is possible to calculate the amount of time it will take for a concentrated substance—a drop of food dye, for instance—to fully diffuse through a volume of water. Just as in this simple experiment, it is also possible to calculate the amount of time required for chemotherapy to be delivered to an entire tumor volume by way of a blood vessel if the penetration length or distance away from the vessel is known. This penetration length depends on the diffusion coefficient for the substance, as well as the rate of uptake of the substance by the live tumor cells, although the environmental conditions are far more complex in the latter case.

3.3.1 Proliferative Index and Apoptotic Index

To gauge the area in the breast tissue affected by DCIS, it is necessary to assess the progression of the tumor in each individual duct. How can this be accomplished at a *single point in time* in just one step? The key is in the ratio of proliferative cells to apoptotic cells in the cross section of the duct. How many cells are proliferating, and how many are in the state of apoptosis?

The fractions of cells in the proliferative and apoptotic states are known as the proliferative index and apoptotic index, respectively. The ratio of these two indices is a key parameter in a cell-scale model of tumor growth that forms the starting point for modeling the area of the entire tumor.

In three tumor modeling papers published between 2011 and 2013, we employed the proliferative and apoptotic indices as key parameters in modeling equations for rendering patient- or experiment-specific predictions of tumor growth in humans and mice and for different cancer types. Cell proliferation is the proliferative division of cells through mitosis, and in the case of a developing tumor, cell proliferation (as well as migration) along the outer edge is significant. Pathologic observations at the cellular scale are made from patient biopsies and can reveal the number of cells in the state of proliferation as a ratio of the total number of proliferating to nonproliferating cells in an observed area. Proliferating cells are identified in pathology through staining with the Ki-67 antibody, which specifically targets the Ki-67 protein found in proliferating cells. In later sections of this chapter, the proliferative index will be referred to as PI, and in some equations the associated rate of mitosis will appear with the symbol λ_M.

Apoptosis is programmed cell death. Similar to the process for determining the percentage of cells in apoptosis, a pathologist can stain the cells obtained from patient biopsies with the cleaved caspase-3 protein antibody and count the number of cells in apoptosis, which, compared with the total number of cells, provides the apoptotic index. The apoptotic index in equations described later in this chapter will be referred to as AI, and the associated rate of apoptosis will appear with the symbol λ_A.

Coupled with observations of proliferation and apoptosis, cellular necrosis (premature cell death due to external factors such as lack of nutrients) is also an important factor in determining the extent of tumor development and progression. The rate of necrosis appears in certain equations with the symbol λ_N.

Measurements of the apoptotic index and proliferative index taken from a single point in time, together with the rate of necrosis, can provide a snapshot of the development of a tumor. These measurements at specific points in time are called *biomarkers*. In much the same way a child's adult height can be predicted by charting the current height and age of the child in months, based on a single observation rather than a series of observations, a tumor's surgical volume can be predicted mathematically, under certain assumptions, based on these cellular-scale parameters taken from immunohistochemistry (IHC) at a single time point.

3.3.2 Diffusion Penetration Length

The final important observation derived from patient histological measurements is the diffusion penetration distance, which appears in the equations with the symbol L. It is the distance, or length, along which cell nutrients such as oxygen can penetrate into a tumor to reach the entire volume of the involved tissue (in the case of these studies, the breast). Pathologists can determine the key parameters, including the tumor volume fraction, f, the nutrient diffusion coefficient, D, and the nutrient uptake rate by tumor cells, λ, which are needed to determine diffusion penetration distance (see below).

3.4 TWO CASE STUDIES

Building on previous scientific work involving mathematical and computational models that have focused primarily on a single breast duct at the microscale [114,171–178], our model offers a multiscale approach, starting small and ending large, at the tissue scale. The work begins at the microscopic scale and creates a cell-scale population dynamics model calibrated with parameter values gained from IHC (i.e., Ki-67 and cleaved caspase-3 staining) and morphometric measurements of parameters such as the duct radius and the thickness of the viable rim of tumor cells within the duct. It is important to recognize that individual breast ducts are microscopic, with a length of approximately 1 mm and a width of only a fraction of this distance, and that the initial calibration work is done by measuring individual cells in breast ducts. When the term *cell scale* is used, it is the scale smaller than a capillary but larger than a molecule, on the order of the size of a single or small group of cells. The work described in this section originates with three separate studies published by our groups. What follows is a summary of these research results, along with extended author information.

3.4.1 Prediction of Surgical Volume

The mathematical pathology approach was first detailed in [9] to describe the characteristics of tumors and predict tumor size through the use of mathematical equations that are calibrated to specific patients' DCIS tumors. The purpose of the study was to show that a very specific diagnosis of tumor size is achievable through measurements that are taken from a single biopsy (for quantifying apoptotic or proliferative indices and diffusion penetration length). These microscopic measurements were used to calibrate a model that, when scaled up to the macroscale, drove the results of a tissue-scale model. We further explain this approach in Figure 3.1; see below for explanations of model parameters.

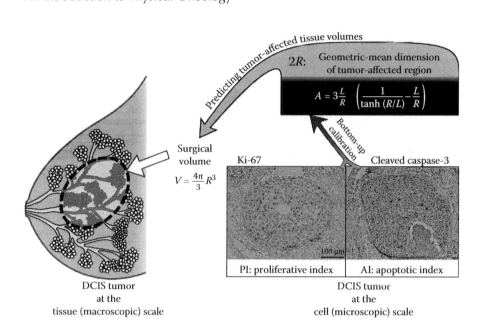

FIGURE 3.1 Prediction of surgical volume from pathology data on an individual patient basis. The measurements on cell scale (i.e., proliferation and apoptotic indices) are averaged over each tumor to calibrate the tissue-scale model. The tissue-scale model is then used to predict the surgical volume. Note that Equation 3.1 of the model depends mechanistically on patient-specific parameters L and A, i.e., nutrient diffusion penetration length in the tumor and ratio of tumor cell death to proliferation, both of which can be obtained from pathology data. (Reproduced from Edgerton, M. E. et al., *Anal. Cell. Pathol. (Amst.)*, 34(5), 247–263, 2011.)

Briefly, results from the numerical experiment were very encouraging, with predictions of tumor size and surgical volume being far more accurate when based on patient-specific biomarkers than were the predictions of the same tumor's surgical volume based on mammography or histologic "grade."

Modeling surgical volume is not necessarily designed to reproduce the exact shape and dimensions of the tumor, but rather to elucidate the extent of the volume of tissue that must be surgically removed in order to (1) increase patient survival and (2) decrease the likelihood that a second or third surgery may be required to completely remove all malignant cells. This less than precise goal in predicting surgical volume involves averaging the physical properties of the tumor.

If early detection reveals the presence of DCIS, we assume that the tumors have already passed their initial fast-growth two- to three-month phase and have reached a size relatively close to their final volume due to the static pressure balance between tumor cell proliferation and tumor cell death. The tumors in DCIS grow to a point where their growth is arrested due to the lack of viable cells capable of proliferating; at this point, nearly all the cells in the core of the duct have undergone lysis (death), with their decayed leftover material contributing to the necrotic core of the duct and accompanying microcalcifications. Hence, there is a balance between tumor cell proliferation and tumor cell death, and the corresponding ratio of tumor cell proliferation to apoptosis, along with the "extent of diffusion of cell nutrients" (diffusion penetration length). This balance reveals the mechanism behind which the mathematics can model the tumor's growth and predict—when the models are scaled up to the tissue scale—the final geographic volume of the tumor, based on one measurement at a single time point.

The analytic solution of the model [179] results in the following "master equation":

$$A = 3 \times \frac{L}{R} \times \left(\frac{1}{\tanh(R/L)} - \frac{L}{R} \right) \tag{3.1}$$

This equation must be calibrated from the results of IHC, which determines cell proliferation and death. Because cell-scale measurements are problematic in modeling tissue-scale tumors, our approach is to input cell-scale measurements into this equation (the cell-scale population model mentioned earlier), which bridges the gap between cell-scale measurements and the tissue-scale model.

IHC, as described earlier, involves staining the cells in the proliferative and apoptotic state, and provides the measurements that drive the proliferative and apoptotic index values. In Equation 3.1, A, the patient-specific ratio of cell apoptosis to proliferation rates averaged over the multitude of ducts within the surgical volume, can be derived from pathology measurements taken on specific patients' tissue. Additionally, morphometric measurements therein provide the values for the diffusion penetration distance, L. When L and A are known, determining the value of R, the geometric-mean tumor surgical radius, is simply a matter of mathematics. In this way, this equation uses the cell-scale values for calibrating the tissue-scale continuum model predicting surgical volume.

The ratio of cell apoptosis to proliferation rates (A) is generated as $A = \lambda_A/\lambda_M$, while L is the product of $L = f^{1/2} \cdot (D/\lambda)^{1/2}$ [179], where the maximum mass growth rate constant by mitosis in ducts is λ_M, the analogous death rate constant is λ_A, f is the tumor volume fraction, and λ is the rate at which nutrients are taken up by cells in the ducts. These values are all calculated by their average values throughout the patient surgical volume, with data being collected manually by pathologists performing measurements and counting cells appearing under different stains in samples collected from patients' resected tumors.

A more complete understanding of the values driving the cell-scale model can be gained by reading the original literature. Table 3.1 summarizes the key parameters of the model. For the sake of clarity, we will explain the process by which the variable A is determined as a ratio of tumor cell apoptosis and mitosis. The mitosis rate, which is the rate of programmed cell proliferation, is represented as λ_M, which can be found in the equation $\lambda_M \cdot \langle \sigma \rangle = PI/\tau_p$, where $\langle \sigma \rangle$ is the averaged nutrient concentration; likewise, the tumor cell apoptosis rate is $\lambda_A = AI/\tau_A$.

What information is gained as a result of these cell-scale measurements and the corresponding values produced by determining the rate of proliferation and cell death? Because of tumor heterogeneity, our most important parameter is the geometric-mean tumor surgical radius, R, which is entered into the basic equation for determining volume (in this case, the surgical volume), that is, $V = 4\pi R^3/3$. How does the value V differ from data generated by

TABLE 3.1   Biophysical Model Parameters

Parameter	Biophysical Meaning
A	Ratio of cell apoptosis to proliferation rates
AI, PI	Apoptotic and proliferative indices
D, λ	Nutrient diffusion coefficient and uptake rate by tumor cells
F	Tumor volume fraction
L	Nutrient diffusion penetration length across tumor surgical volume
R	Geometric-mean tumor surgical radius
R_{duct}, T	Duct radius and viable rim thickness
λ_A, λ_M	Tumor cell apoptosis and mitosis rates
σ, σ_H	Nutrient concentration and concentration at the perinecrotic boundary
τ_A, τ_P	Cell apoptosis and proliferation times

Source: Reproduced from Edgerton, M. E. et al., *Anal. Cell. Pathol. (Amst.)*, 34(5), 247–263, 2011.

mammography? How is mathematical pathology more useful than current imaging? Our research indicates that surgical volumes determined by radiology and tumor grade are largely inaccurate when compared with surgical results. Our study involved examining 17 excised DCIS tumors, and we found that mammography overestimated the tumor size in 10 cases and underestimated the tumor size in 7 cases. *The correlation between mammography, nuclear grade, and the final observed tumor size is poor.* In one example (specimen 19), mammography determined that the geometric mean was approximately 6 cm, but upon pathological examination after excision, it was determined that the geometric mean was actually only 2 cm. A contrasting specimen (18L) had a 1 cm geometric mean by mammography but was actually greater than 3 cm by pathologic analysis after excision (Figure 3.2). However, *the correlation between the sizes observed after surgery and those predicted by the model was close* (Figure 3.3).

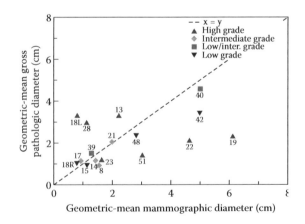

FIGURE 3.2   Comparison between the geometric-mean tumor diameters from mammograms and from pathology analysis. Data were obtained from excised DCIS tumors (de-identified case numbers are reported). Mammography overestimates the tumor size in 10 cases and underestimates the other 7. (Reproduced from Edgerton, M. E. et al., *Anal. Cell. Pathol. (Amst.)*, 34(5), 247–263, 2011.)

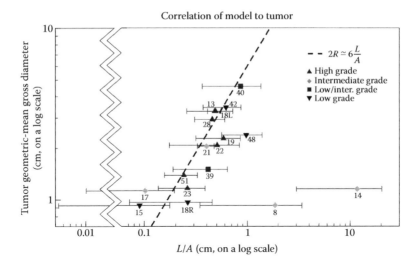

Correlation of model to tumor

FIGURE 3.3   Correlation of tumor size with the death-to-proliferation ratio parameter. Tumor geometric-mean diameters $2R$ (dashed) vs. L/A predicted by the model compared with the corresponding pathology measurements. In contrast, grades based on histopathology are clearly poor predictors of tumor size. Thus, L/A is a novel, more accurate biomarker of tumor size. Note that $2R \simeq 6\dfrac{L}{A}$ is a valid approximation of Equation 3.1 for large tumors where R exceeds the value of L. Data were obtained from the 17 excised tumors (symbols, with de-identified case numbers). (Reproduced from Edgerton, M. E. et al., *Anal. Cell. Pathol. (Amst.)*, 34(5), 247–263, 2011.)

In agreement with [180], the apoptotic and proliferative indices show similar trends in their relationships with tumor grade (Figure 3.4A). This confirms that net proliferation (i.e., ratio of PI to AI), and thus tumor size, *has a weak correlation with histologic grade*. To clarify, the histologic grade of a tumor was assigned by a pathologist with low, low–intermediate, intermediate, or high grade; the lower the grade, the more slowly the cancer cells grow, and the better the patient prognosis. Moreover, viable rim thickness of tumor in ducts, and thus L (the nutrient diffusion penetration length), decreases as a function of histological grade (Figure 3.4B). It can be interpreted that more proliferative, high-grade tumors result in tightly packed cellular structures, and thus are likely to hamper oxygen and nutrient diffusion. Lack of oxygen in high-grade tumors would drive hypoxia-inducible factors and cell migration, which may lead to penetration of the ductal wall and infiltration of the tumor into the breast stroma, which is also in line with data that have been reported previously [114,174].

The simulation of DCIS growth confirms published findings that 80% of DCIS tumors have reached 95% of the extent of their growth at the time of diagnosis, and confirms the likelihood that these tumors are static by the time of mammographic imaging. The implications of the accuracy of these predictions is that with standard mammogram and pathologic specimens, physicians should be able to accurately predict the surgical volume of DCIS tumors mathematically, resulting in a lower chance of additional required surgeries. The study also confirms that tumor grade or tumor dimensions from mammography

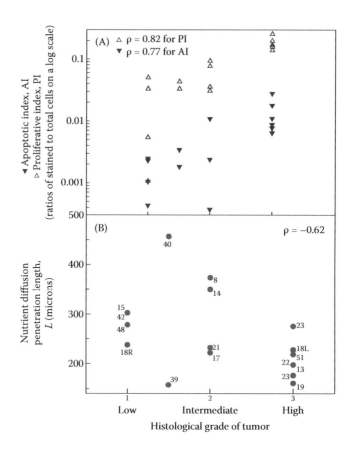

FIGURE 3.4 Correlations of IHC and morphometric measurements with nuclear grade. (A) Average apoptotic (AI) and proliferative (PI) indices for each tumor as increasing functions of (modified Black's) nuclear grade (one overstained AI tumor and one zero AI tumor were excluded). $\rho = 0.82$ for PI and $\rho = 0.77$ for AI indicate that both indices have similar correlations with tumor grade. (B) Nutrient diffusion penetration length, L, from average measured viable rim thickness in each tumor's ducts is a decreasing function of nuclear grade. (Reproduced from Edgerton, M. E. et al., *Anal. Cell. Pathol. (Amst.)*, 34(5), 247–263, 2011.)

are inconsistent with actual tumor volume. Together, *this study represents a proof of principle that it is possible to incorporate a mathematical modeling step within current clinical practice to aid in and improve surgical planning by estimating the surgical volume and the outcome of surgery before treatment.* Since IHC and morphometric measurements can be performed on patient-specific breast biopsies, the clinical value of this mathematical pathology approach is that the prediction and resulting surgical planning can be tailored for that particular patient. The practice will lead to less subjective analysis of tumors and improved surgical treatment efficacy through individualized treatment design.

3.4.2 Prediction of Tumor Growth

Building on the positive results of the paper we discussed in the previous section [9], we developed an even more detailed model to predict tumor progression starting at the microscopic scale using a lattice-free agent-based modeling (ABM) approach [181]. This model is the first to account for how cellular calcification influences tumor progression, and is fully constrained to patient-specific clinical data easily obtained from histopathology. The study raised key questions about DCIS that, when answered, could further assist in surgical planning and understanding tumor growth: How do we describe the physical process of DCIS from its genesis to its detectable form? Can mathematical modeling help us interpret how DCIS develops in patients? How can modeling assist in understanding tumor growth and in surgical planning? How do the microcalcifications detected in mammographic imaging relate to tumor morphology? If IHC is performed on only a single biopsy sample, can these data be used to calibrate patient-specific models that predict tumor growth over time?

Briefly, in the ABM (see [181] for details), cells (agents) are modeled as physical objects that exchange and respond to adhesive, repulsive, and motile forces that determine their motion. Cell–cell and cell–microenvironment interaction mechanics are modeled using potential functions that account for finite interaction distances, uncertainty in cell morphology and position, and interaction between cells of variable sizes and types. Each cell is initialized with a phenotypic state (see Figure 3.5), and phenotypic transitions are governed by exponentially distributed random variables that depend on the cell's internal state and the local microenvironment.

We then couple the agents with the microenvironment by introducing field variables for key microenvironmental components (such as oxygen, growth factors, and

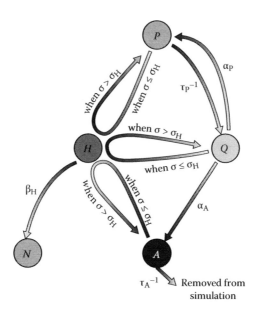

FIGURE 3.5 Phenotypic transition algorithm. Cell states include quiescence, Q; proliferation, P; apoptosis, A; hypoxia, H; and necrosis, N. Specifically, quiescent cells can become proliferative (P) or apoptotic (A), nonnecrotic cells become hypoxic (H) when oxygen σ drops below a threshold value σ_H, and hypoxic cells can recover to their previous state or become necrotic (N). (Reproduced with permission from Macklin, P. et al., *J. Theor. Biol.*, 301, 122–140, 2012.)

extracellular matrix [ECM]) that are governed by reaction–diffusion equations. In the model, cells alter the evolution of the environmental variables, and these variables also affect the cells' behavior. For example, at the macroscopic scale, oxygen transport is modeled by

$$\frac{\partial \sigma}{\partial t} = \nabla \cdot (D \nabla \sigma) - \lambda \sigma \tag{3.2}$$

where σ is oxygen, D is its diffusion constant, and λ is the uptake or decay rate. Here, we briefly explain how to calculate λ in the model. Suppose that in a small neighborhood of a position $\mathbf{x}$, there are three types of cells, tumor cells, host cells, and stroma (background environment) cells, and those cells occupy fractions f_t, f_h, and f_b, respectively, where $f_t + f_h + f_b = 1$. Also, viable tumor cells consume oxygen at an uptake rate λ_t, host cells at a rate λ_h, and elsewhere oxygen decays at a low (background) rate λ_b. Then, $\lambda(\mathbf{x})$ is determined by averaging the uptake rates with weighting according to the local tissue composition, which is consistent with the uptake rate model [182]:

$$\lambda(\mathbf{x}) = f_t \lambda_t + f_h \lambda_h + f_b \lambda_b \tag{3.3}$$

We applied the modeling technique described in [9] to further model the development of microcalcifications: the necrotic core of DCIS tumors. A key addition to the model is modeling the movement of cells within a simulated approximately 1 mm section of a duct, based on biomechanical forces and cell morphology. The most important aspect for model calibration is how to quantify patient-specific model parameters using clinically accessible histopathology data. We briefly discuss our method and focus on DCIS here, although the technique can be applied more generally to tumors with clearly visible viable rims. Table 3.2 summarizes the patient-specific parameters of the model.

We obtain these parameter values either from histopathology data analysis or by solving the system of partial differential equations to steady state (see the original publication for

TABLE 3.2   Patient-Specific Parameters

Parameter	Biophysical Meaning
R_C	Cell radius
R_N	Cell nuclear radius
σ_B	Oxygen value of the basement membrane
$\langle \sigma \rangle$	Mean oxygen level in viable rim
$\langle \alpha_p \rangle$	Mean transition rate for quiescence to proliferation (cell proliferation rate)
α_P^{-1}	Mean waiting time before the transition of quiescence to proliferation when $\sigma = 1$
α_A	Transition rate for quiescence to apoptosis (cell apoptosis rate)
S	Cell spacing
c_{cca}, c_{cba}	Cell–cell and cell–basement membrane adhesive force coefficients

Source: Reproduced with permission from Macklin, P. et al., *J. Theor. Biol.*, 301, 122–140, 2012.

details). Two important parameters, cell proliferation and apoptosis rates, are obtained as follows:

$$\langle\alpha_P\rangle=\left(\frac{1}{\tau_P}(PI+PI^2)-\frac{1}{\tau_A}AI\cdot PI\right)\Big/(1-AI-PI),$$

$$\langle\alpha_A\rangle=\left(\frac{1}{\tau_A}(AI-AI^2)+\frac{1}{\tau_P}AI\cdot PI\right)\Big/(1-AI-PI). \tag{3.4}$$

To verify the calibration, we ran a simulation for 30 days of simulated time, and we computed the simulated AI and PI, mean viable rim thickness, and viable rim cell density at 1-hour increments for the last 15 days of simulated time. Figure 3.6 shows the mean and standard deviation of the simulation results compared with the patient data. Because all the results for each parameter are within the range of patient variation, the calibration can be considered successful.

With parameter values quantified for each individual patient, the model can simulate the growth dynamics of DCIS (see Figure 3.7 for one simulation example). Results from this predictive model of tumor development correspond to clinical data describing the

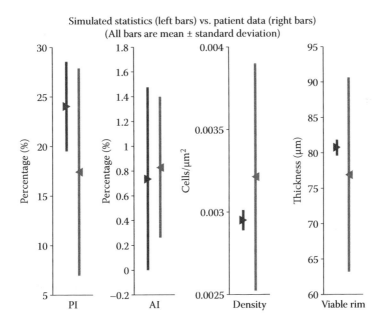

FIGURE 3.6 Model parameter verification. Comparison of the simulated (left bars) and patient (right bars) data for PI (column 1), AI (column 2), cell density (column 3), and viable rim thickness (column 4) over the last 15 days of our simulation. Since the bars overlap for each parameter, and the simulated means (left triangles) fall well within the patient variation for each parameter as well, it is concluded that the calibrated model matches the calibration data within tolerances. (Reproduced with permission from Macklin, P. et al., *J. Theor. Biol.*, 301, 122–140, 2012.)

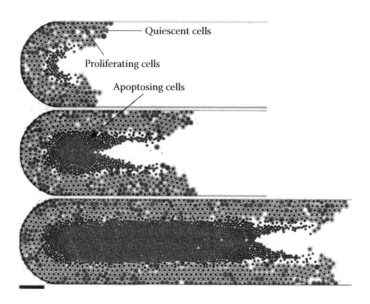

FIGURE 3.7 Simulation results of the DCIS ABM in a 1 mm length of duct. Small dark circles, cell nuclei; dark circles in the center of the duct, necrotic cells. The shade of the necrotic debris indicates the level of calcification. Simulated times (from top to bottom): 11, 24, and 45 days. Bar: 100 μm. (Reproduced with permission from Macklin, P. et al., *J. Theor. Biol.*, 301, 122–140, 2012.)

growth of DCIS, and the model confirms the hypothesis that duct diameter is an important factor in tumor growth, as the rapidly dividing cells tend to lyse toward the core of the duct. We determined that

$$x_V = 20.52 + e^{6.085 - 0.02584 R_{duct}} \text{ μm/day} \tag{3.5}$$

where x_V is tumor advance rate and R_{duct} is the radius of the duct. It can be observed that the wider the duct, the slower the growth rate axially along the duct, while the narrower the duct, the faster the tumor progresses axially (see Figure 3.8). The measurements taken at the microscopic scale can be placed in a model that predicts tumor growth at a macroscopic scale. The model predicts tumor growth of 7.5–10.2 mm/year (at the macroscale) based on measurements of the tumor's apoptosis-to-proliferation ratio, which can only be measured at the microscopic scale using histologic data.

We then use the ABM to predict the quantitative relationship between the mammographic (x_V) and pathologic (x_C) tumor sizes and obtain

$$x_V \approx 0.4203 + 1.117 x_C \text{ mm} \tag{3.6}$$

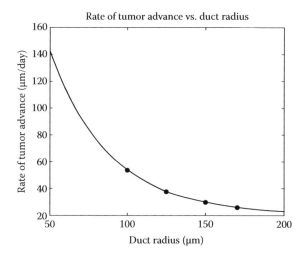

FIGURE 3.8   An inverse correlation is observed between the duct radius and rate of tumor advance. For small ducts, little lumen is available for mechanical relaxation, leading to rapid tumor advance; conversely, the growth is slower for larger ducts. Solid curve, plot of Equation 3.5; solid circles, simulation results of the ABM. (Reproduced with permission from Macklin, P. et al., *J. Theor. Biol.*, 301, 122–140, 2012.)

We compare the plot of this equation against our simulated DCIS data (Figure 3.9A, solid circles) and a set of published clinical data by another group (Figure 3.9B, squares) [183]. We find that our model not only correctly predicts a linear correlation between a DCIS tumor's mammographic and pathologic sizes, but also demonstrates an excellent agreement with published clinical data two orders of magnitude larger than our simulation data.

Together, the ABM has provided a number of biological insights into the biology of DCIS. While the mathematics behind the predictions is complex, the models begin with data that are elegant in their simplicity: every portion of data taken from patient pathology can be generated in most cancer centers nationwide—the practice can be replicated just about anywhere cancer is treated, and is the starting point for what drives the models. No extra steps or tests are needed to obtain the necessary parameters that drive the model, making the model easy to apply and convenient to use for physicians.

The most important scientific inquiry conducted in this study involves describing the physical forces acting between cells, such as cell adhesion to other cells, repulsion forces between cells, and the cells' locomotive forces along the basement membrane of the duct. The work done modeling DCIS growth and cell–cell or cell–basement membrane interaction confirms that patient-specific measurements taken at a single time point are useful in predicting tumor growth and tumor volume.

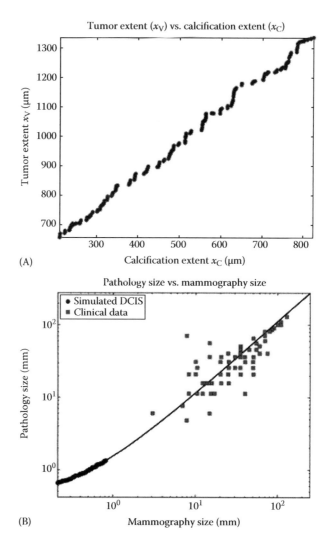

(A)

(B)

FIGURE 3.9 Comparison of mammographic (x_V) and pathologic (x_C) DCIS sizes. (A) A linear correlation between x_C and the actual pathology-measured x_V is found from our ABM simulation results. (B) A linear least-squares fit to our simulation data (solid circles on bottom left) fits a clinical dataset (squares), further demonstrating the predictivity of the ABM model. (Reproduced with permission from Macklin, P. et al., *J. Theor. Biol.*, 301, 122–140, 2012.)

3.5 APPLICATION TO LYMPHOMA GROWTH

In another extension of the original work on modeling DCIS growth based on microscopic measurements of cell death and proliferation, in collaboration with Dr. Sanjiv Gambhir (Stanford University) and Dr. Hermann Frieboes (University of Louisville), the model was applied to a different form of cancer: non-Hodgkin's lymphoma (NHL) [157]. NHL is a disseminated, highly malignant cancer that commonly becomes drug resistant, and in the United States alone, one-third of patients will die from this disease within five years of diagnosis [184–186]. The purpose of this modeling effort is to gain further insight into the tissue-scale effect of molecular-scale mechanisms that drive lymphoma growth, and also to study their association to cell proliferation, death, and physical transport barriers within the tumor tissue.

Measurements of cellular apoptosis and proliferation were generated in mice by way of injections of cancerous cells into the tail vein while the animals were in a state of depleted immunity. We then determined how different phenotypes of cancer cells (specifically, drug sensitive and drug resistant) develop within the tumor or microenvironment and might be affected by chemotherapy. While the previous two studies on tumor development [9,181] were primarily directed at assisting surgery by predicting surgical volume in breast cancer, this study also has applications for the diagnosis of cancer types, as well as planning treatment when chemotherapy is an option in lymphoma. Measurements attained from histology data, when applied to the model, can be used to predict not only tumor size, but also potentially tumor growth dynamics and drug response. The model could inform physicians of the likely effectiveness of a treatment plan and the likelihood of drug resistance in the tumor prior to treatment, thus better informing physicians and allowing them to develop more effective treatment plans.

3.5.1 Investigation of Tumor Heterogeneity in Drug Resistance

NHL in mice is similar enough to NHL in humans to serve as an appropriate analog when designing predictive tools for human treatment applications. From these mouse studies, we have identified five different variables relevant to tumor progression: cell viability, hypoxia in the microenvironment, vascularization of the tumor, proliferation of tumor cells, and cell apoptosis within the tumor. We investigated (and modeled) not only tumor growth, but also the role of a tumor's physical heterogeneity in its development of drug resistance— genetically because of the lymphoma cell type, as well as physically from transport barriers present in the microenvironment. The *in vivo* model (in mice) provided information regarding the genetic role in tumor growth and drug resistance, as well as pathological features of the disease in mice, that can be compared to the human disease. In the model, the tissue microstructure was constructed through the proper choice of parameter values and through biologically justified functional relationships between these parameters, for example, cellular transitions from quiescence to proliferation, depending on oxygen concentration. The model was used to simulate asymmetric tumor evolution and dynamically coupled heterogeneous growth, vascularization, and tissue biomechanics. Figure 3.10 shows our overall approach to constrain the model parameters on both cell and tumor scales.

In order to understand the role of physical heterogeneity in the development of drug resistance, we focused on two types of lymphoma cells: drug-sensitive *Eμ-myc Arf–/–* cells and drug-resistant *Eμ-myc p53–/–* cells. We also chose the Eμ-myc transgenic mouse model (which expresses the Myc oncogene in the B-cell compartment) as our *in vivo* model because it captures genetic and pathological features of the human disease; thus, drug-resistant and drug-sensitive tumors can be directly compared [187,188]. In our experiments, the lymphoma tumors in the mice were removed after a time period of 21 days and prepared for IHC. A total of five sets of histology sections were obtained, and the sections in each set

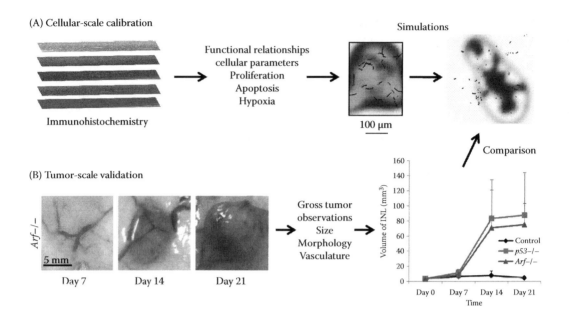

FIGURE 3.10 Integrated experimental and computational modeling strategy. (A) Obtain functional relationships involving cell-scale parameters such as proliferation, apoptosis, and hypoxia from experimental data. For example, the density of viable tissue can be defined as a function of vascularization (see the panel in the middle: dark colors, highest density; light colors, lowest density; segments, vessels). These functional relationships and parameter values (measured experimentally) are then used as input to the model for simulating lymphoma growth (see the sample simulated tumor cross section at the far right). (B) Validate the model results with tumor-scale lymphoma data obtained from macroscopic imaging of a lymph node in live mice; these data include the size, morphology, and vasculature of the tumor. (Reproduced from Frieboes, H. B. et al., *PLoS Comput. Biol.*, 9(3), e1003008, 2013.)

were stained for cell viability (hemotoxylin and eosin [H&E]), hypoxia (hypoxia-inducible factor-1 [HIF-1a]), proliferation (Ki-67), apoptosis (caspase-3), and vascularization (CD31); see Figure 3.11 for measurements of cell proliferation and apoptosis for both cell lines. After staining the five-layer sections of the resected tumor, pathologists measured the cell-scale parameter values of proliferation, apoptosis, viability, vascularization, and hypoxia. We then counted individual cells in multiple layers of the tumor specimens, examining all the individual variables mentioned above at the cellular level. These cell-scale measurements from pathology were input into the mathematical model to make predictions of tumor growth.

After calibrating the lymphoma model with the cell-scale measurements, we validated the simulated tissue-scale lymphoma size with *in vivo* macroscopic and intravital imaging data at the tissue scale. Model predictions of tumor geometric mean paired with the actual tumor scale *in vivo* from macroscopic observations of tumor growth are shown in Figure 3.12. It was demonstrated that the model accurately predicted the final geometric mean of each tumor type after 21 days of growth, with only slight inaccuracy in the model's prediction of the acceleration of tumor growth during the initial fast-growth stage of 10–14 days.

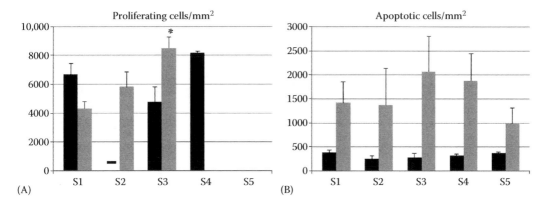

FIGURE 3.11 Histological data analysis of the lymphoma. (A) Proliferating cells per area and (B) apoptotic cells per area for the five sections measured (S1–S5, positioned from top to bottom) for the two cell lines, drug-sensitive *Eμ-myc Arf–/–* (black) and drug-resistant *Eμ-myc p53–/–* (gray). All error bars represent standard deviation from at least $n = 3$ measurements in each section; an asterisk indicates statistical significance ($p < 0.05$) determined by Student's t-test with $\alpha = 0.05$. There is higher vascularization in the center, higher hypoxic density on the periphery, and higher overall apoptotic density in *Eμ-myc p53–/–* compared with *Eμ-myc Arf–/–*. (Reproduced from Frieboes, H. B. et al., *PLoS Comput. Biol.*, 9(3), e1003008, 2013.)

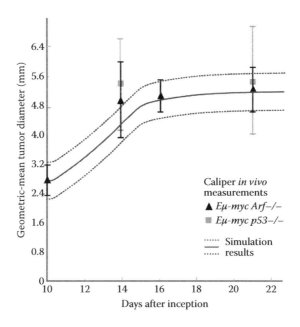

FIGURE 3.12 Comparison of model predictions with experimental lymphoma growth data. Mean tumor diameter (solid line) predicted by the model, with predicted variation (dashed lines) based on uncertainty in the measured oxygen diffusion distance, falls within the range of values measured for the tumor growth observed *in vivo* (denoted by the triangles and squares with vertical error bars). (Reproduced from Frieboes, H. B. et al., *PLoS Comput. Biol.*, 9(3), e1003008, 2013.)

The model then described lymphoma growth greatly beyond the original lymph node size, which is consistent with *in vivo* observations. Observations also indicated that there was no statistical difference between the size of each type of lymphoma tumor.

We then compared the simulated vasculature to independent *in vivo* intravital microscopy data of a *Eµ-myc p53–/–* tumor in the same animal over time (Figure 3.13). We also observed that if we increased the lymphoma cell population at the beginning, the homogeneous distribution of cell substrates (e.g., oxygen and nutrients) changed accordingly, leading to diffusion gradients of these substances, which in turn impacted the lymphoma cell viability. If we began with a heterogeneous cell population within the tumor (as observed experimentally in *Eµ-myc Arf–/–* cells near the tumor periphery), the model showed that the diffusion gradients would not be as pronounced. Conversely, if the cell viability is

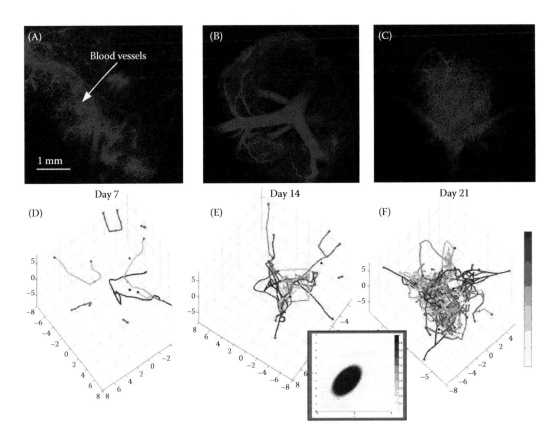

FIGURE 3.13 Simulation of vasculature and angiogenesis in the lymph node tumor. (A–C) Observations in an *Eµ-myc p53–/–* tumor in living mice using intravital microscopy. These data provide information to qualitatively compare the vessel formation with the computer model predictions (D–F, correspondingly). Modeling the diffusion of cell substrates (e.g., oxygen and cell nutrients) within the tumor allows for prediction of the spatial distribution of lymphoma cells as their viability is modulated by access to the oxygen and nutrients diffusing from the vasculature into the surrounding tissue. Inset: One vessel cross section. (Reproduced from Frieboes, H. B. et al., *PLoS Comput. Biol.*, 9(3), e1003008, 2013.)

higher near the center of the tumor, then the gradients are predicted to be steeper and more uniform.

This modeling work [157] extends our mathematical pathology approach's usefulness in predicting tumor growth, as it applied *in vivo*—that is, in the mice—data, including immunohistological data gathered from the specimens, to modeling. Additionally, the study investigated the growth of two types of lymphoma cells: *Eμ-myc Arf–/–* and *Eμ-myc p53–/–*, which exhibit differing sensitivities to the chemotherapy drugs doxorubicin (DOX) and cyclophosphamide.

Overall, this process yielded a lymphoma simulator as an initial step to study detailed tumor progression and provide further insight into drug resistance, and ultimately may provide a tool to design better personalized treatments for NHL. Since the cell-scale measurements used for calibration are different from those at the tissue scale used for verification, this methodology enables the model to bridge from the cell to the tumor scale to calculate tumor growth and hypothesize the associated mechanisms predictively, that is, without resorting to fitting to the experimental data. This process quantitatively links cellular phenotype to tumor tissue-scale behaviors, and may serve to highlight the importance of physical heterogeneity and interactions in the tumor microenvironment when evaluating chemotherapeutic agents, in addition to consideration of chemoprotective effects, such as cell-specific phenotypic properties and cell–cell and cell–ECM adhesion [38].

3.5.2 Implications of Diffusion Barriers and Chemotherapy Design

One goal of this study was to simulate diffusion barriers within the lymph node tumor over time. The barriers against drug diffusion stem from the tumor density of viable cells and vascularization profiles; consequently, these values impact levels of oxygen and nutrients, driving the tumor to attain a heterogeneous final state conditioned by variable diffusion gradients of these substances within the tumor. Differences in cell viability across the tumor, from the center to the outer rim (i.e., tumor structure heterogeneity), and the distances and densities of tumor through which chemotherapy drugs like DOX and cyclophosphamide must penetrate will impact diffusion gradients. The research also indicates differences in drug diffusion between the two types of tumors studied, which when applied to humans could help design chemotherapy treatments tailored to the individual tumor phenotype. We intend to pursue the hypothesis that genetic differences, such as those tested in this study, will account for differences in responses to chemotherapy, according to measurable variables taken from patient biopsies: proliferation, apoptosis, and hypoxia. Furthermore, this study encourages further investigation into how modeling can contribute to patient-specific treatment plans, as it includes predictions of how diffusion barriers might impact drug diffusion; see [158] for our modeling effort in this endeavor, which will be further discussed in Chapter 5.

3.6 CONCLUSIONS

Successes in predicting tumor volume and growth rates based on measurements taken from microscopic pathology provide justification for mathematical modeling of cancer as a promising step in developing treatment plans—especially when planning for surgery such as breast conservation surgery. The mechanisms by which these data can be gained and integrated into patient treatment plans exist in almost any cancer treatment facility. Therefore, with further research, we hope to show that we can accurately predict clinically relevant outcomes so that these models can assist physicians and patients in making treatment decisions. In Chapter 4, we discuss how macroscopic imaging, such as contrast-enhanced computed tomography scans, can also be used to derive data to drive models predicting the fraction of tumor killed by chemotherapy in various types of cancer.

Mathematical Modeling of Drug Response

With Joseph D. Butner

CURRENT PROCESSES FOR CREATING cancer treatment plans are primarily based on empirical data from clinical trials, consensus expert panel guidelines, and limited laboratory-based testing, but physicians still face major challenges in successful before-treatment prediction of the effectiveness of any particular treatment plan for any given patient. Work done in physical oncology not only can help improve treatment outcomes (as examined in Chapter 3), but also could benefit many cancer patients if incorporated into the creation of treatment plans. Here, we will examine three published papers that provide evidence for improving cancer treatment plan efficacy through the inclusion of mathematical modeling techniques, based on our capability to consider the multiple scales of the human body (cellular, tissue, and organ) when predicting patient-specific treatment outcomes. More specifically, we first developed a mathematical model that is capable of predicting the response of cancer cells to any drug concentration *in vitro* based on the uptake rate of drug by the cells [10]. Next, we used a mathematical model to predict breast cancer growth inhibition *in vitro* and *in vivo*, in order to demonstrate the inhibitory effects of tissue-scale diffusion barriers on drug delivery [189]. Finally, we developed a more general model that includes a spatial variable at the tissue scale to predict tumor response, leading to the discovery of how the effectiveness of chemotherapy is dependent on the vasculature of the tumor microenvironment [154].

4.1 CURRENT STATE: CELLULAR, TISSUE, AND ORGAN LEVELS

Currently, cancer treatment plans are limited by a significant information gap during planning and development: lack of consideration of physical barriers and dynamics at the tissue scale. Current treatment plans are informed by results from cellular (genetic)-level and organ- or whole body-level tests, but ignore the tissue level, which falls in between. The cellular level contains factors that can be observed in the cancer cells, including genetic mutations and expression of certain proteins that may be related to specific drugs or prognosis. The tissue level considers the whole tumor and often includes many of the culprits of drug resistance mentioned in Chapter 2: pH, oxygen, and nutrient availability or limitations; vasculature; interstitial fluid pressure; cell-to-cell charge distribution; and cell density. Including information from tissue-scale experiments or individual patients' tissue-scale parameters in treatment regimen design could impact and improve cancer treatment outcomes, reduce costs, and lower the unnecessary side effects often incurred during chemotherapy.

Treating cancer is costly. In 2012, the average three- to four-month chemotherapy regimen cost between $20,000 and $26,000 (Avalere Health [190]), in addition to impacting the patient's health and quality of life. An effective uniformity in treatment plans (as relative to the patient's needs) could save money, time, and unnecessary side effects from treatment. Organizations like the National Comprehensive Cancer Network (NCCN) produce evidence-based treatment guidelines that take into account findings from the newest publications and therapies. These guidelines provide recommendations to physicians about which specific variables should be taken into consideration during the treatment process. These variables include the type of anticancer drug used, and its particular dosage, frequency, and duration of the treatment cycle; however, these values are primarily determined from *in vitro* monolayer experiments or clinical studies, both of which do not take into account explicit contributions from the tissue scale.

According to the current state of cancer chemotherapeutic treatment, if a patient shows signs of cancer, a physician will perform a physical exam and order the necessary tests for further evaluation. Once the presence of cancer is confirmed by imaging or laboratory tests, a biopsy (if deemed necessary or beneficial) can be done to determine the specific cancer type. The combination of exams, laboratory data, and imaging data also help to determine the stage of the cancer. The clinical team designing the plan will take into account the patient's age and overall health, and the stage, location, and type of cancer to determine the most effective treatment strategy [191]. However, these guidelines do not recommend the usage of any information about tissue-scale parameters (i.e., vasculature, pH, presence of hypoxia, and cell density).

The three main strategies for treating cancer include one or a combination of anticancer drugs (chemotherapy, immunotherapy, hormone therapy, or targeted therapy), surgical resection, or radiation therapy. Chemotherapy is given for the purposes of shrinking cancer before surgery, preventing cancer recurrence after surgery, or simply killing or preventing growth of existing cancer in the absence of any surgical resection [192]. Despite current guidelines, there is still a significant degree of variation in treatment outcomes between different patients due to the absence of individual treatment optimization based on the conditions and needs of the specific patient. While recommended guidelines from multiple organizations exist, which include recent studies and treatment options, these guidelines are best practices for a general population—not individual patients whose tumors will respond differently to a particular treatment. Without considering the biophysical characteristics of an individual patient's tumor and how it will respond to a specific treatment, treatment plans will continue to fail to maximize effectiveness, while minimizing side effects, treatment duration, and costs.

In order to further examine how cancer treatment regimens are currently designed and implemented, we will pull an example directly from the NCCN website's guidelines, specifically in our case for treating breast cancer [193]. If the physician finds potential signs of cancer during a physical exam, he or she will order blood tests and liver function tests to check for evidence that the cancer has spread to the bones or liver, and is thus affecting blood counts or creating abnormal levels of chemicals in the liver. Imaging tests—usually a mammogram—can often identify the location and size of the tumor. A biopsy or cancer cell test will then be performed by the clinician and analyzed by a pathologist. For the case of breast cancer, the pathologist usually looks for the prevalence of hormone receptors and HER2-neu receptors. Treatment plans are based off of these tests. If high levels of hormone or HER2-neu receptors are present, physicians prescribe specific drugs that inhibit growth hormones from stimulating cancer growth or inhibit HER2-neu receptors. Depending on the observed location, size, presence of particular receptors, presence of lymph nodes, and specific type of cancer, physicians may consider chemotherapy an option. For example, chemotherapy is considered essential for higher-risk breast cancers (invasive lobular carcinoma, invasive ductal carcinoma, metaplastic carcinoma, and mixed carcinoma) that are hormone and HER2 negative, and have tumor sizes greater than 1 cm. Chemotherapy is downgraded to only a potential treatment if, under the same hormonal and receptor conditions, the tumor is only 0.51–1.0 cm with no lymph node tumors, or is 0.5 cm or smaller where lymph tumors are also present.

There are many scenarios where chemotherapy is suggested or may potentially be beneficial. In these situations, if the medical team chooses to use any form of anticancer drug, they must decide which drug or combination of drugs, and the concentration, duration, and frequency to administer the drug. Whenever possible, physicians administer drugs that are proven effective against the type of cancer present. In the example above, the drug or drugs chosen will depend on the type of breast cancer, as well as the presence or absence of certain receptors on the tumor. There is no consideration of the specific physical microenvironment of an individual tumor, and our data support the notion that this may help to tailor the treatment to a specific patient by selecting drugs or drug formulations that would be predicted to have better delivery to the cancer cells, and therefore more treatment effect. Instead, many drugs in clinical trials are based on the results of *in vitro* (petri dish) experiments or overly simplistic animal studies. In the petri dish, a specific drug is tested on a single layer of a particular strain of cancer cells. In the animal models, the tumor development occurs in anatomic locations where transport barriers are not as difficult to overcome (such as subcutaneous tumors) as in the actual disease, or other components of the cancer are not present (such as the immune system). Thus, many drugs that are effective in laboratory experiments do not translate well to patients in terms of improving their outcomes.

4.2 THE NEED FOR INCORPORATION OF TISSUE-SCALE MODELING

While many drugs have proven effective in the lab in *in vitro* and *in vivo* experiments, why are the same results often not observed in human patients? And why do physicians prescribe varying treatment plans for similar patients? A possible reason for the discrepancies in cancer drug administration, and the current ineffectiveness of anticancer drugs, is due to the fact that only two major scales are taken into account when determining a treatment regimen. Cancer drugs currently in use are proven to be effective on a cellular scale—killing particular cancer cells in a monolayer—and their mechanisms of action at this scale are well understood. Specific drugs have even been designed that are able to overcome potential genetic or cell-level barriers to more effectively kill cancer cells.

Drugs like doxorubicin (DOX) are designed and proven to work on a cellular level. These drugs passively diffuse or use the cell's own mechanisms to pass through the membrane into the cell, and usually into the nucleus to disrupt the cell's DNA replication or other essential cellular functions, ultimately leading to apoptosis. DOX specifically inhibits topoisomerase II, resulting in breaks in the DNA backbone during replication and thus triggering apoptosis. All these effects on killing cancer cells have been perfectly proven on the cellular level.

Clinical trials also provide valuable information on the effectiveness of drugs at an organ level, where the observable effectiveness can be observed in patients. However, neither of these scales provides an adequate explanation for the prevalence of drug resistance and the variation in effectiveness of drugs across patient cohorts, or the poor results of overall treatment when compared with monolayer or clinical experiments. The need to include the tissue scale is clear, which is why we have been focusing on understanding, modeling, and using tissue-scale parameters to predict treatment outcomes for individual patients with more accuracy, thus enabling the creation of more effective and individualized treatment plans.

4.3 MODELING CONCEPT

Cancer researchers must study the tumor's physical properties at all levels, from the cellular to the tissue and organ scales, to successfully describe and evaluate a particular treatment's ability to successfully administer drugs to the cancer cells at a high enough concentration to effectively kill or shrink the tumor. The following studies were conducted by our groups with a focus on addressing the tissue-scale gap. Understanding the mechanisms of drug delivery to the tumor, its diffusion to—and through—the tumor, and potential barriers to delivery is the focal point of this research.

As evidenced in these modeling studies, we have shown that tissue-scale factors significantly affect a drug's ability to kill cancer. Focusing on this scale allowed us to observe the tumor microenvironment's impact on drug delivery, which revealed the presence of diffusion barriers, which prevent adequate concentrations of drug from reaching the tumor cells. We found not only that tissue-scale barriers impact drug delivery and effectiveness, but also that diffusion barriers may be a main cause for drug resistance in many tumors. If the sources of drug resistance can be determined, researchers and physicians can create strategies to overcome them, thus drastically improving treatment outcomes.

We have studied the impact of the tissue scale on drug delivery in various types of cancer, including non-Hodgkin's lymphoma, breast cancer, colorectal cancer, and pancreatic cancer. We found that particular biomarkers provide valuable information on a drug's potential effectiveness. These biomarkers include average blood vessel radius, length of drug diffusion from the vessel, prevalence of vasculature, hypoxia, stromal density, and T-cell population. We also designed and optimized a general mathematical model that is capable of predicting the fraction of tumor cells killed (f_{kill}) from an anticancer drug–based therapy (primarily chemotherapy in our case) based on measurements of patient-specific parameters that can be measured before treatment from routine, noninvasive scans. Applicable to multiple cancer types, this model provides the ability, in the early stages of the treatment planning process, to predict a certain treatment's potential effectiveness. The model also provides an understanding of the mechanisms for drug resistance and lays the foundation for future *in vivo* studies to test possible strategies to overcome drug resistance based on these tissue-scale barriers.

This research may have immediate clinical relevance, not only because of its ability to predict treatment outcomes, but also because it lays the groundwork for personalized medicine. On a tissue scale, each individual tumor is different, with its own microenvironment and biophysical characteristics. If these tissue-scale characteristics can be quantified and accounted for, patient-specific treatment plans can be composed to specifically overcome the barriers present in an individual tumor. Quantification of tissue-scale characteristics would allow physicians to follow a general guideline for treatment with the inclusion of modifications for an individual patient's disease state and treatment needs, and make treatment decisions with much more information on how effective different treatments may be, depending on the particular tumor.

4.3.1 Modeling of Drug Delivery to Predict Cytotoxicity for MDR versus Drug-Sensitive Cell Lines *In Vitro*

In collaboration with C. Jeffrey Brinker (University of New Mexico), we set out to model the delivery of DOX in liver hepatocellular carcinoma (HCC) tumor cells *in vitro* [10]. We sought to mechanistically understand the differences between the delivery of conventional free drug and nanoparticle-based delivery, as well as the differences in response of multi-drug-resistant (MDR) cell lines versus drug-sensitive cell lines. This study aids in understanding the tumor's physical response to therapy and provides a clearer picture of which parameters need to be identified and overcome to defeat drug resistance at multiple scales. We succeeded in developing a mathematical model that predicts tumor cell response *in vitro* based on measurable cellular parameters. This study gave direction to further research on barriers to drug delivery *in vivo* and established the ideal of individualized medicine in human patients based on predictions made from mathematical models.

We predicted that the dynamics of the cells' response to treatment could be understood and predicted using only the experimental measurements of the cell death rate and uptake rate. The mass conservation of drug is expressed as [194,195]

$$\frac{d\sigma}{dt} = -\lambda \cdot n(t) \cdot \sigma, \quad \sigma(t=0) = \sigma_0 \tag{4.1}$$

where t is time, $\sigma(t)$ is the local drug concentration, λ is the specific rate of uptake of drug by the cells (inverse time per unit concentration of cells), and $n(t)$ is the concentration of cells. The drug is assumed to be at initial concentration σ_0. The conservation of cells is given by

$$\frac{dn}{dt} = -\Lambda_A \cdot n, \quad n(t=0) = n_0 \tag{4.2}$$

where $\Lambda_A(t)$ is the death rate of the cells (inverse time), and the initial concentration of cells is n_0. Although Michaelis–Menten kinetics is commonly assumed to describe the cell death rate Λ_A [196–199], here *we assume that this rate is a function of the total amount of drug taken up by the cells up to time* t, as described by

$$\Lambda_A(t) = f\left(-\int_0^t \lambda \cdot n(\tau) \cdot \sigma(\tau) \cdot d\tau \right) = \lambda_A \cdot \left(\sigma_0 - \sigma(t) \right) \tag{4.3}$$

The argument of the integral is the total rate of uptake of drug by the cells at any time. We have assumed the function f to be linear, and thus the specific cell death rate λ_A to be constant. Equation 4.2 becomes

$$\frac{dn}{dt} = -\lambda_A \cdot n \cdot \left(\sigma_0 - \sigma(t) \right) \tag{4.4}$$

We emphasize that this model (Equations 4.1 and 4.4) describes the dynamics of the viable cell population as a function of drug concentration and the history of drug uptake by the cells. This model is unique in that it couples cell and drug dynamics, which are dependent on the drug's ability to kill at different concentrations, as well as the cell's ability to take up drug over time. Model parameters can be measured directly from monolayer experiments with both MDR and drug-sensitive cell lines.

To test the model's prediction ability, we treated MDR HCC cells with free DOX and DOX-loaded protocells* [149,200], each at two different concentrations for four hours. Using these two data points, the model was "calibrated" with the hope of being able to predict cell kill at any drug concentration with conventional or protocell delivery. We also tested the generality of the model's predictive ability by performing continuous time exposure experiments to MDR and drug-sensitive HCC cells using other drug types (such as cisplatin and 5-fluorouracil) and other cell lines, such as breast cancer and pancreatic cancer cell lines (we encourage the reader to refer to the original article [10] for more details on these model results).

---

* A protocell is a form of nanocarrier made of a silica core with a high-surface-area matrix. High levels of drug are stored in the core, and the protocell is surrounded by a lipid bilayer that is designed to have very specific interactions with the environment, allowing the targeting and binding of specific cells.

We investigated the effect of the duration of cell exposure to DOX on cell death (Figure 4.1) by using experimental data corresponding to delivery for four hours to MDR HCC cell lines via both free DOX and DOX-loaded protocells at two initial DOX concentrations, σ_0. We applied the values of the parameters $\lambda \cdot n_0$ and λ_A (previously calculated from an independent continuous DOX exposure experiment *in vitro*) in this model, because these parameters are intrinsic properties of the cells, and thus are independent of the time delivery protocol according to our assumptions. We found that the results predicted by the mathematical model agreed with the data corresponding to four-hour drug exposure (coefficients of determination $R^2 = 0.988$ and 0.982 for the free DOX and DOX-loaded four-hour exposure curves at an initial DOX concentration $\sigma_0 = 92.5$ nM, and $R^2 = 0.995$ and 0.973 for $\sigma_0 = 298$ nM), thus validating the model assumptions and equations. When drug

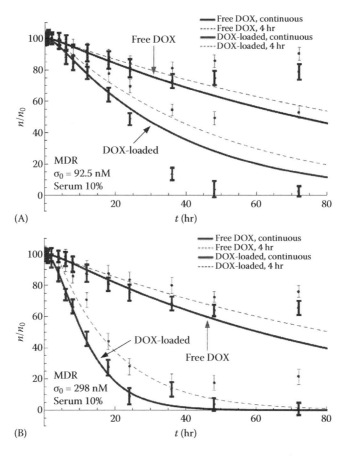

FIGURE 4.1 Continuous delivery of DOX over time (thick symbols and solid lines) leads to higher longtime death rates than delivery for four hours (thin symbols and dashed lines). Percentage of viable cells n/n_0 vs. time t (symbols with standard deviation) for free DOX and DOX-loaded protocells delivered for 4 hours and continuously for 72 hours to MDR HCC cell lines. Initial DOX concentrations: $\sigma_0 = 0.0925$ µM (A) and 0.298 µM (B). The mathematical model (lines), with parameters calibrated from the continuous-drug exposure experiments, is predictive of the viability corresponding to the four-hour exposure experiments. (Reproduced with permission from Pascal, J. et al., *ACS Nano*, 7(12), 11174–11182, 2013.)

is delivered for 4 hours only, cells begin to regrow with a doubling time of mitosis of about 25–32 hours, while simultaneously the time-dependent kill effect of the drug declines; thus, we do not expect the model as calibrated to accurately predict cell viability values for times longer than this doubling time. Consistent with the mathematical model, continuous delivery leads to higher total uptake, that is, $\sigma_0 - \sigma_\infty > \sigma_0 - \sigma(4\text{ hour})$, because $\sigma(4\text{ hour}) > \sigma_\infty$, and thus higher death rates and total death are achieved. This result is consistent with the observation that the uptake timescale 2 hour $< (\lambda \cdot n_0)^{-1} < 15$ hour, and thus after four hours, the drug uptake process is typically far from being completed.

We then predicted theoretical dose–response curves* spanning the full range of drug concentrations by numerically integrating the model up to $t = 24$ hours. Figure 4.2 shows a comparison of the theoretical and experimental dose–response curves, that is, a fraction of viable cells n/n_0 at $t = 24$ hours versus initial drug concentration σ_0, for free DOX and DOX-loaded protocells. The resultant coefficients of determination R^2 were 0.987, 0.991, and 0.996 for the MDR cell line with free DOX, DOX-loaded protocells, and the parental cell line with free DOX, respectively. *The ability to accurately predict the entire dose–response curves from parameter fits based on results of the in vitro experiments using only two initial concentrations, despite the nonlinear nature of drug response and without incorporating complicated kinetic arguments to describe cell death, provides further support to the mathematical model assumptions.* For the case where DOX-loaded protocell and free DOX delivery are compared, the results imply that the delivery of DOX using protocells is more efficient (because of higher cellular uptake rates) and affords a kill equal to the kill achieved by delivery of free DOX at a higher total concentration or at a longer drug exposure time (equivalence).

The initial concentration of drug available for uptake by the cells was found to be the most important variable. The higher this initial concentration, the greater the uptake rate, long-term cell death rate, and total drug uptake by the cells. Protocells were able to achieve greater death rates because of their ability to deliver larger drug dosages to inside the cell. This is largely due to their ability to carry large doses and to avoid efflux pumps, which eliminate free DOX from the cell. More importantly, the model results revealed that cancer cell death rates are an intrinsic property that can be predicted based on the time integral of drug exposure. The longer cancer cells are exposed to drugs, and the higher the initial concentration of drug available for uptake, the greater the amount of cell death. This research also proved that drug resistance can be overcome if cancer cells are provided sufficient drug for a long enough period of time. Ultimately, the success of this study encouraged future studies exploring the mechanisms of successful drug delivery and the possibility of overcoming drug resistance.

We here note that even if chemotherapy drugs are able to kill drug-resistant cell lines *in vitro*, it may be possible for unaccounted-for tissue-scale barriers to drug delivery *in vivo* to prevent drug-resistant cells from receiving the adequate amount of drug necessary to achieve maximum cell kill. This hypothesis led us to explore the possible mechanisms for

---

* Dose–response curves are graphs plotting the quantity of viable cells on the *y*-axis and the drug concentration used on the *x*-axis. These curves provide information on the relationship between drug dosage concentrations and the resultant cell kill efficacy.

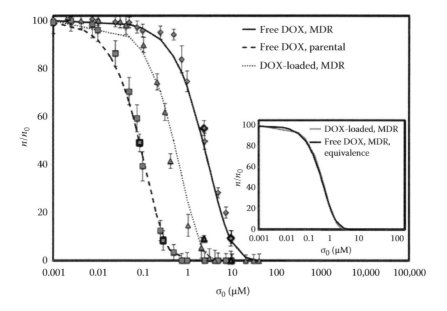

FIGURE 4.2 Comparison of dose–response curves predicted by the model and observed cell kill obtained from experimental cytotoxicity data (symbols with standard deviation). Free DOX: squares, parental, i.e., drug-sensitive cell line; diamonds, MDR, DOX-loaded protocells: triangles. The obtained R^2 was >0.9 for each of the three cases, i.e., (MDR, free DOX), (MDR, DOX-loaded protocells), and (parental, free DOX). Inset: Equivalence between free DOX and DOX-loaded protocell dose–response is demonstrated. (Reproduced with permission from Pascal, J. et al., *ACS Nano*, 7(12), 11174–11182, 2013.)

barriers to drug delivery *in vivo* and how they can be overcome. This particular experiment employed protocells as an example of a potential solution to overcoming the barriers of drug delivery, as protocells possess the ability to deliver high concentrations to cancer cells for longer periods of time than free drug delivery, thus maximizing cell kill.

In this study, we were able to use a mathematical model to predict the outcome of drug delivery at different doses to drug-sensitive and MDR cells of multiple cancers types using multiple drugs. In particular, using data points gained from *in vitro* experiments with only two initial concentrations for four hours, the model was able to predict dose–response curves for each cell type, delivery method, and drug tested. This study brought to light the potential of a mathematical model in predicting cell kill *in vivo* and even in human patients. If predicting cell kill in humans is possible, it could have significant clinical relevance in creating the most effective treatment plans for individual patients.

4.3.2 Diffusion Barriers: Vital Patient-Specific Chemotherapy Inhibitors

Cancer cells—even drug-resistant lines—can be completely killed in monolayer experiments if a large enough initial concentration of drug is available for uptake (although due to side effects, this high dosage may not always be feasible in a patient). The success of anticancer drugs in monolayer experiments is well documented, but a need to further understand why many cancer therapies have limited efficacy in patients is still present.

Hence, we attempted to address this need by modeling the diffusion through and uptake by breast cancer cells of cytotoxic immune molecules *in vitro* and *in vivo* [189]. Breast cancer remains the leading cause of death for women ages 20–59 worldwide, and current treatments have only proven effective in initial disease control [201].

T cells play a significant role in the immunosurveillance and destruction of cancer cells. The tumoricidal potential of γδ T cells was derived from the observation of preferential expansion and infiltration of γδ T cells in various types of tumors [191]. However, since this initial observation, limited success has been obtained clinically in attempts to translate this knowledge into effective immunotherapies [202]. One of the major impediments is limitations on the infiltration of a sufficient number of such cytotoxic cells within the tumor. Indeed, a very small number of γδ T cells were found within breast tumor tissues in patients with various stages of disease [203].

γδ T cells recognize antigens directly, without any requirement for antigen processing and presentation [204,205]. As a major mechanism of γδ T-cell-mediated growth control, interferon-γ (IFN-γ) (a small cytotoxic molecule secreted by γδ T cells) induces antiproliferative and pro-apoptotic effects on many types of cancer cells [206]. Upon activation, human γδ T cells secret IFN-γ in a dose-dependent manner, and this is correlated with the γδ T-cell-mediated cytotoxicity [207,208]. Since γδ T cells represent only a small subset (2%–5%) of the total T-cell population [203], a major impediment for γδ T-cell-mediated therapy is the delivery and infiltration of a sufficient amount of γδ T cells (and thus IFN-γ molecules) to the tumor.

To model the diffusion and update of IFN-γ over time across the tissue region where γδ T cells and cancer cells coexist, we assume that these processes are quasi-steady compared with the proliferation and death processes. The concentration of IFN-γ across the tissue region (a mixture of γδ T cells and cancer cells) is described using the following partial differential equation:

$$0 = D \cdot \nabla^2 \sigma - \lambda \cdot \sigma \tag{4.5}$$

where σ represents the interstitial concentration of IFN-γ, λ is the uptake rate of IFN-γ by the cancer cells, and D is the interstitial diffusion constant of IFN-γ.

We performed one *in vitro* and two *in vivo* experiments. Human γδ T cells were extracted and isolated from the periphery of human blood. In the *in vitro* monolayer experiment, the T cells were exposed to the breast cancer cell line SKBR7. The γδ T cells successfully inhibited breast cancer cell survival and proliferation, with the maximum inhibition found at 30 γδ T cells per tumor cell.

In the first *in vivo* experiment, mice were simultaneously injected with γδ T cells and SKBR7 cells at ratios of 15:1 and 30:1. After four weeks, the tumors were removed, and their size and the number of apoptotic cells present were measured. As expected, the tumors coinjected with γδ T cells at 15:1 and 30:1 ratios were reduced to 51.6% and 33.09% of the average control tumor volume. Next, γδ T cells were injected into mice after two weeks of tumor growth. One hundred million γδ T cells were injected, and four weeks later, the ratio of γδ T cells to tumor cells was estimated to be between 0.004 and 0.067, much smaller than the previous experiment. Tumor size was also evaluated, but in this case, the treated tumors were only 22.1% smaller than the controls.

We then compared the mathematical model predictions of tumor cell kill with these *in vitro* and *in vivo* measurements of γδ T-cell-mediated breast cancer growth inhibition; this is shown in Figure 4.3. We found that the optimal kill ratio of 30:1 γδ T cells to cancer cells found *in vitro* is unachievable in human patients because of adverse cytotoxic effects. For example, in the mice the ratio of 15:1 γδ T cells to cancer cells was more effective, likely because at higher ratios (e.g., 30:1), γδ T cells began to kill each other. Ratios of 0.004 and 0.067 γδ T cells to cancer cells (estimated in the second *in vivo* study four weeks after injection) are much more likely to be found in humans.

With the model, we predicted that therapeutic effectiveness could be improved by accounting for the diffusion of cytotoxic molecules within the tumor tissue scale. While γδ T-cell immunotherapy is effective *in vitro* and in extremely high concentrations in mice, it has proven to be an ineffective treatment in human cancer patients. Low efficacy in patients is likely due to T-cell scarcity and barriers to diffusion. We suggest that diffusion at the tissue scale is critical in determining the success of tumor growth inhibition and lysis. This suggests that clinicians should measure the biophysical transport barriers (e.g., oxygen supply, vasculature, pH, and tumor cell density) identified before treatment. If this is possible, the barriers can be overcome or normalized so that more anticancer agents, in this case γδ T cells, can be delivered to the tumor cells.

As previously mentioned, success *in vitro* does not translate to success in treating actual cancer patients. This presents a major problem when creating treatment plans based on results from monolayer experiments, especially when determining drug concentration values. This study clarified the necessity of identifying the underlying mechanisms preventing successful delivery of anticancer drugs.

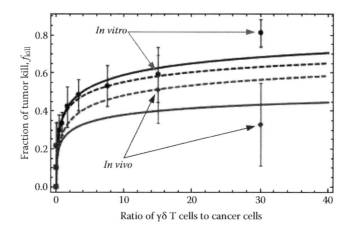

FIGURE 4.3 Fraction of tumor kill vs. ratio of γδ T cells to cancer cells (φ). Experiments *in vitro* (dark circles) and *in vivo* (diamonds, φ = 15, 30; circles, 0.004 < φ < 0.067; squares, φ ≈ 0.037). Mathematical model (top solid curve, least-squares fit to the *in vitro* data; R^2 = 0.99 and *p*-value = 0.0002; bottom solid curve, least-squares fit to the *in vivo* data; R^2 = 0.84 and *p*-value = 0.0037; dashed curves, fits without outliers). (Reproduced from Das, H. et al., *PLoS One*, 8(4), e61398, 2013.)

4.3.3 Generalized Model for Predicting Tumor Response to Drug Treatment

We sought to improve our model to include not only time, but also spatial factors. We believe that a model that is able to account for diffusion on a tissue scale, in addition to cellular and genetic influences, would bridge a gap in understanding and predicting tumor response to anticancer therapies. We extended the time-dependent model [10] by incorporating spatial dependence to describe perfusion and diffusion heterogeneities. The governing equations for drug concentration $\sigma(\mathbf{x}, t)$ and the volume fraction of tumor cells $\varphi(\mathbf{x}, t)$ are

$$\frac{\partial \sigma}{\partial t} = D\nabla^2\sigma - \lambda_u\varphi\sigma \tag{4.6}$$

$$\frac{\partial \varphi}{\partial t} = -\lambda_u\lambda_k\varphi(\mathbf{x},t)\int_0^t \sigma(\mathbf{x},\tau)\varphi(\mathbf{x},\tau)d\tau \tag{4.7}$$

where D is the diffusivity of the drug, λ_u is the per-volume cellular uptake rate of drug, and λ_k is the death rate of tumor cells per unit cumulative drug concentration. Because drug diffusion occurs much faster than the process of cell death, Equation 4.6 can be solved at the quasi-steady state (i.e., $\partial\sigma/\partial t \cong 0$). Thus, without the time derivative in Equation 4.6, the solution $\sigma(\mathbf{x}, t)$ is independent of initial conditions; for boundary conditions, we set a drug concentration σ_0 at the blood vessel wall. For a cylindrically symmetric domain surrounding a blood vessel (Figure 4.4), the boundary conditions can be set to

$$\sigma(r = r_b, t) = \sigma_0 \tag{4.8}$$

and

$$\mathbf{n} \cdot \nabla\sigma|_{\mathbf{x}\to\infty} \to 0 \tag{4.9}$$

where r denotes the radial position from the center of the cylinder, and r_b represents the blood vessel radius; the second boundary condition reflects that the far-field drug concentration asymptotes to zero at $\mathbf{x} = $ infinity. Furthermore, Equation 4.7 has no spatial derivatives, and thus only requires the initial conditions for $\varphi(\mathbf{x}, t)$, which we set to

$$\varphi(\mathbf{x}, t = 0) = \varphi_0 \tag{4.10}$$

that is, a homogeneous initial tumor volume fraction. As detailed below, with this simple revision, this generalized model allows us to examine not only successive (conventional) bolus chemotherapy (characterized by a time-varying intravenous drug concentration σ_0 according to a specific dosing and timing regimen), but also drug release through loaded nanocarriers, where drugs are released at a nearly constant rate over a certain time interval, approximated here by a constant σ_0.

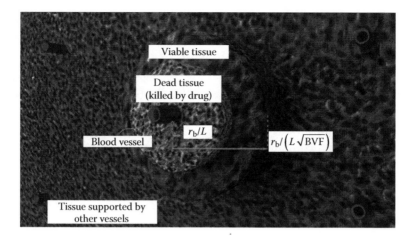

FIGURE 4.4 Illustration of transport-based hypothesis. A blood vessel supplies substrates (e.g., oxygen) to the cylindrical tissue volume surrounding the vessel by diffusion. We hypothesize that at each position inside the tissue, the substrate supply is supported exclusively by the closest blood vessel. Thus, the influenced tissue surrounding a vessel can be estimated to be between a cylinder of radius $r_b/(L\ \mathrm{BVF}^{1/2})$ in dimensionless form and the vessel itself with dimensionless radius r_b/L. Theoretically, chemotherapeutic drugs delivered by a blood vessel kill the tissues immediately adjacent to the vessel, leaving some viable tissues on the far end (away from the vessel). *Here, we propose that through drug-loaded nanocarriers that can accumulate within tumors and continuously release drugs for a longer time (e.g., lasting several cell cycles), the drugs can penetrate further into the surrounding tissue volume (relative to free drug delivery), and thus achieve a higher tumor killing ratio.* (Reproduced from Wang, Z. et al., *PLoS Comput. Biol.*, 12(6), e1004969, 2016.)

By integrating the viable tumor volume fraction at each time point over the cylindrical tissue domain surrounding a blood vessel and affected by the drug diffusion, we calculate f_{kill} as the ratio of the killed tumor volume to the total initial tumor volume,

$$f_{\mathrm{kill}}(t)=1-\dfrac{\dfrac{2}{\varphi_0 L^2}\displaystyle\int_{r_b/L}^{r_b/(L\sqrt{\mathrm{BVF}})}\varphi(r,t)r\,dr}{\left(\dfrac{1}{\mathrm{BVF}}-1\right)\dfrac{r_b^2}{L^2}} \tag{4.11}$$

as a function of parameters: r_b (blood vessel radius), blood volume fraction (BVF), and $L=\sqrt{D/(\varphi_0\lambda_u)}$ (the effective diffusion penetration length of the drug). As shown in Figure 4.4, the model domain is comprised of the space between two concentric cylinders, specifically the blood vessel and the surrounding tissue it supplies. The inner cylinder has a radius r_b/L in dimensionless units, representing the blood vessel at the center of the domain. With the hypothesis that the substrate supply for any spot in a tissue is supported by the closest blood vessel, we estimate $r_b/(L\ \mathrm{BVF}^{1/2})$ to be the dimensionless radius of the influenced tissue volume of the vessel, where BVF < 1. The influenced tissue volume refers to a specific region of tissue that relies on this blood vessel for supply of oxygen and other essential chemicals.

We performed a set of simulations with the generalized time- and space-dependent model in a cylindrically symmetric domain surrounding a blood vessel. Model parameter values were set as follows: $L = 155.06$ μm and $r_b = 15.83$ μm (based on experimentally measured values in resected patient colorectal metastasis to liver tumors), giving $r_b/L = 0.102$; we also set BVF = 10^{-2} as the reference value for the simulations, as measured BVF values ranged approximately from 10^{-3} to 10^{-1}. We then ran the model for 10 drug-induced apoptotic cycles. To examine the impact of each parameter on f_{kill}, we further simulated nine parameter variation combinations, using three r_b/L values, that is, 0.05, 0.1, and 0.5, paired with three BVF values, that is, 0.005, 0.01, and 0.05. Figure 4.5 shows the numerical results of the model. In the presence of a boundary condition σ_0 at the vessel wall ($r = r_b$), successive cell layers next to the blood vessel die out (Figure 4.5A) due to cell-death-induced enhancement of drug penetration (Figure 4.5B), in turn leading to accelerated cell

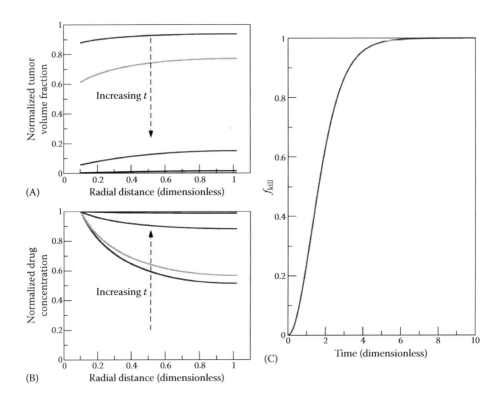

FIGURE 4.5 Numerical simulations of the general integrodifferential model in a cylindrically symmetric domain. As cell kill ensues over several cell cycles, (A) successive cell layers next to the blood vessel ($r = r_b$) die out, i.e., tumor volume fraction φ decreases (top to bottom: $t = 0.5, 1, 3, 5$, and 10); (B) local drug concentration σ increases due to an enhancement of drug penetration (bottom to top: $t = 0.5, 1, 3, 5$, and 10); and (C) cell kill accelerates farther from the vessel and deep into the tumor. Input parameters: $r_b/L = 0.102$ and BVF = 0.01. The duration of the entire simulation was 10 $(\lambda_k \lambda_u \varphi_0 \sigma_0)^{1/2}$, where time unit is a characteristic cell apoptosis time. Drug concentration and tumor volume fraction were normalized by their initial values, and radial distance by the diffusion penetration distance L. The fraction of tumor kill f_{kill} is calculated from Equation 4.11. (Reproduced from Wang, Z. et al., *PLoS Comput. Biol.*, 12(6), e1004969, 2016.)

kill (Figure 4.5C). As cell kill occurs, tumor volume fraction φ decreases, leading to an increase in local drug concentration σ (because dead cells no longer take up drugs), and thus accelerating cell kill in the locations farther away from the vessel and deeper into the tumor.

Figure 4.6A shows the temporal evolution curves of f_{kill} calculated from Equation 4.11 by varying the parameters r_b/L and BVF, representing conditions where drug-loaded nanocarriers are employed, as well as the estimates using our simplified "bolus" model. The results indicate that the bolus kill ratios are readily achieved by drug-loaded nanocarriers

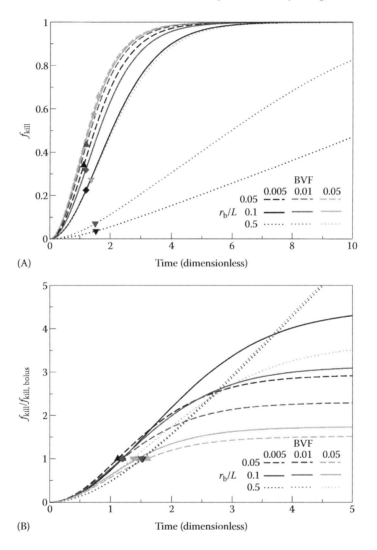

FIGURE 4.6   Drug-loaded nanocarriers lead to cell kill enhancement over bolus delivery. (A) Time–evolution curves of chemotherapeutic efficacy f_{kill} (Equation 4.11) of nanocarriers releasing drug compared with the estimated efficacy (symbols) of conventional chemotherapy, for parameter values $r_b/L = 0.05$ (dashed curves, upward-pointing triangles), 0.1 (solid curves, diamonds), and 0.5 (dotted curves, downward-pointing triangles), paired with BVF = 0.005, 0.01, and 0.05, respectively. (B) Same as (A), but normalized to the corresponding bolus values of tumor kill, $f_{kill,\ bolus}$. (Reproduced from Wang, Z. et al., *PLoS Comput. Biol.*, 12(6), e1004969, 2016.)

after one or two cell cycles. To estimate the benefits of drug release by the nanocarriers over a longer period of time, we normalized the f_{kill} curves using their corresponding bolus f_{kill} values (Figure 4.6B). *The results suggest that we may achieve two- to fourfold of the bolus killing ratios if the drug release from nanocarrier administration lasts for three or four apoptotic cycles.* However, for large BVF values (representing highly vascularized tumors), cell kill effects from both methods of delivery are roughly equivalent. This is expected because the majority of the tumor cells are killed within just one or two apoptotic cell cycles; a 50% increase in tumor kill is nevertheless expected from loaded nanocarriers releasing over a longer period of time. This suggests an alternative strategy to improve chemotherapeutic efficacy by promoting or normalizing angiogenesis at the target site before administrating chemotherapy drugs [126,209–211], or by promoting perfusion by other means, such as mild hyperthermia [212], both of which would lead to an increase in BVF.

We then performed a parameter perturbation analysis to determine the impact of each parameter on overall f_{kill}. For each of the two parameters, r_b/L and BVF, we created 11 variations, through a range of ±50% of the parameter's standard value and with a 10.0% variation interval. Note that a 50% variation has been assumed as reasonable in systems modeling analysis [213–215]. Accordingly, this generated a total of (11 × 11 =) 121 parameter variation pairs, covering a wider range of parameter space. Figure 4.7 shows the simulation results of f_{kill} from changing BVF and r_b/L, at $t = 3$ dimensionless time units. We find that smaller r_b/L and greater BVF values lead to larger f_{kill}, and thus increased treatment effects. This is because in our simulations, the domain size (i.e., the radius of the outer cylinder) was determined by $r_b/(L \cdot BVF^{1/2})$. Thus, a larger r_b/L or a smaller BVF represented a larger tissue volume relying on the modeled blood vessel for drug transport, and hence would require a longer time to achieve the same f_{kill}.

From all these analyses, we found that the smaller the BVF value, the more likely the tissue is to rely on a single vessel for drug, due to the scarcity of vessels. This scarcity

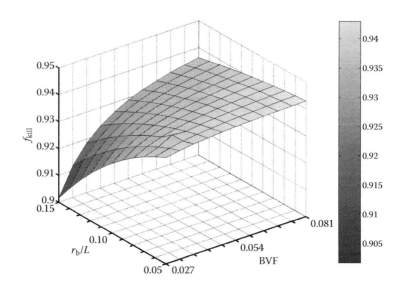

FIGURE 4.7 Effects of combinatorial change in BVF and r_b/L on f_{kill}.

of vasculature results in a longer required exposure to the drug to kill the tumor cells. Depending on the BVF value, the drug administration time required to adequately kill the tumor cells changes dramatically. Tumors with large BVF values have a plentiful supply of blood; therefore, the drug can reach the entire tissue in one apoptotic cycle, which is two to four times less time than if the tumor has a small BVF value.

We also found that larger r_b/L values were associated with larger vessels, likely resulting in a greater amount of drug transported further into the tissue, because more drug is available in the large blood source. If physicians were informed of this value, they could promote angiogenesis or vasodilation during treatment to ensure that a sufficient amount of drug could reach the entire tumor. Using these three parameters alone, this research clearly demonstrated that the efficacy of treatment is greatly dependent on tissue-scale factors. One may envision extending our generalized model to other therapeutic strategies that aim to improve efficacy through enhanced drug delivery by increasing BVF. These strategies include, but are not limited to, vascular normalization [209], metronomic dosing [216], and stromal targeting [217,218]. Each of these physical sciences–based therapeutic strategies could be optimized through our predictive mathematical model.

4.3.4 Application to Prediction of Drug Treatment via Nanocarriers

A series of pilot studies have tested this hypothesis using experimental tumor models for delivering drugs via multistage vectors (MSVs) [219–221]; MSVs are nested particles (with smaller particles nested inside larger ones), particularly designed to lodge the larger, outer particles in tumor vasculature in order to release the smaller, nested nanoparticles over an extended period of time (e.g., several weeks). Here, we further simplify the model presented in the previous section to quantitatively formulate the hypothesis of drug delivery via loaded nanocarriers, as histology data are not always available for determining values (for parameters BFV, r_b, and L). Together with our experimental collaborators, Dr. Haifa Shen and Dr. Mauro Ferrari (Methodist Hospital Research Institute), we then used the model to predict the tumor response to different forms of drug delivery methods before the start of treatment. Model predictions were further validated using experiments on a breast cancer mouse model *in vivo*.

Thus, we developed an alternative form of f_{kill} as a function of another set of experimental parameters, the values of which can be obtained from *in vivo* cytotocixity experiments. By substituting Equation 4.6 into Equation 4.7, we have

$$\frac{\partial \varphi}{\partial t} = -\lambda_k D \int_0^t \nabla^2 \sigma \, d\tau \tag{4.12}$$

Integrating φ in Equaiton 4.12 over the total tissue volume V, we obtain the changing rate of the tumor volume:

$$\frac{d}{dt} \int_V \varphi \delta V = -\lambda_k D \int_0^t \left(\int_V \nabla^2 \sigma \delta V \right) d\tau \tag{4.13}$$

We denote the tumor volume as V_T; by the convergence theorem, Equation 4.13 can be rewritten as

$$\frac{dV_T}{dt} = -\lambda_k D \int_0^t \left(\int_{\partial V} -\frac{\partial \sigma}{\partial n} \delta a \right) d\tau \tag{4.14}$$

where ∂V represents the boundaries of the total tissue volume in question here, and $\partial \sigma / \partial n$ is the flux of the drugs across the boundaries. It is safe to hypothesize that the flux of drugs becomes negligible at the tissue boundaries far away from the blood vessel, and hence the only contribution in the boundary integral we consider is the flux at the boundaries next to the blood vessel. For simplicity, we define

$$F \equiv D \int_{\partial V} -\frac{\partial \sigma}{\partial n} \delta a \cong D \int_{r_b} -\frac{\partial \sigma}{\partial n} \delta a \tag{4.15}$$

We further hypothesize that the rate of change of flux for the first several days under MSV drug delivery conditions is approximately zero: $dF/dt = 0$ (i.e., this initial time period is too short for F to change significantly). Hence, F is constant, and we have

$$\frac{dV_T}{dt} = -\lambda_k \int_0^t F \, d\tau = -\lambda_k F t \tag{4.16}$$

which leads to

$$V_T = V_{T,0} - \frac{1}{2} \lambda_k F t^2 \tag{4.17}$$

or equivalently,

$$\frac{V_T}{V_{T,0}} = 1 - \frac{\lambda_k F t^2}{2 V_{T,0}} \tag{4.18}$$

In fact, $1 - (V_T/V_{T,0})$ is exactly our definition of f_{kill}, that is, the fraction of tumor volume killed from chemotherapy. Hence, we obtain a new mathematical formula for calculating the amount of f_{kill} through the delivery method of loaded nanocarriers:

$$f_{kill} = \frac{F \lambda_k}{2 V_{T,0}} t^2 \tag{4.19}$$

Note that there is a quadratic increase in f_{kill} with time, which is consistent, as previously observed *in vitro* [10].

4.3.4.1 In Vivo *Experiments*

In order to compare the efficacy of free versus nanopartricle drug delivery *in vivo*, a series of experiments were conducted in BALB/c mice at the Houston Methodist Research Institute. Mice were injected in their inguinal mammary fat pads with 5×10^4 green florescent protein (GFP)–labeled murine 4T1 cells (day 0). After two weeks, tumors were observed to have grown to a volume $V_{T,0} = 100$–200 mm^3. Mice were sorted into four random groups of 10 and intravenously administered one of four treatment protocols: (1) phosphate-buffered saline (PBS) administered twice weekly, control; (2) free DOX 3 mg/kg administered twice weekly; (3) 1.0 μm porous silicon nanoparticles (PSPs); and (4) 2.6 μm PSP, both PSP sizes loaded with chemotherapy drug (6 mg/kg) administered once weekly. All tumor volumes were measured on days 14, 17, 21, 25, 28, and 31. Tumors were excised subsequent to mouse sacrifice via CO_2 asphyxiation on day 31. For all treatment groups, measured excised tumor volumes were normalized to average volumes of control (PBS) tumors, and for each tumor, f_{kill} was calculated as 1 minus the normalized tumor volume.

4.3.4.2 Model Predictions

Time evolution of f_{kill} (i.e., Equation 4.19) for the three treatment groups is seen to remain approximately constant after rapidly growing during the first three days of treatment (Figure 4.8, $f_{kill} = 0$ at treatment day 0). Here, the quadratic time dependence seen in Equation 4.19 can be related to the rapid growth of the dead cell fraction in the first three days. The measured f_{kill} of the PSP delivered drug is seen to be roughly three times the kill from the free

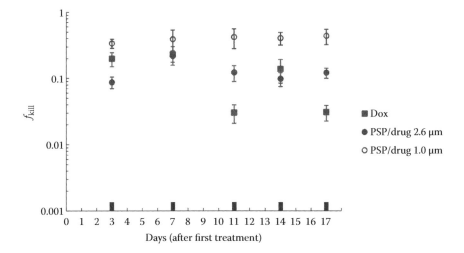

FIGURE 4.8 Testing the efficacy of drug-loaded nanocarriers in mice. Comparison of fraction of tumor killed measured across three different treatment BALB/c mice groups ($n = 10$ per group) over a period of 17 days (from day 14 to day 31 after 4T1 tumor cell inoculation) showing a roughly threefold increase in kill from nanovectored drug vs. free drug. At each time point, tumor volume measurements from the three drug treatment groups were first normalized to the measurement from the control (PBS) group (no drug treatment), and then to the initial tumor volume for each group; f_{kill} was then calculated as (1 − normalized tumor volume). (Reproduced from Wang, Z. et al., *PLoS Comput. Biol.*, 12(6), e1004969, 2016.)

DOX, at an f_{kill} value of approximately 0.5. *This is in good agreement with model predictions of a roughly two to four times increase in tumor kill in the case of PSP delivery over free drug* (as dependent on parameter values; see Figure 4.6B). As per the experimental protocol, the total amount of drug released by the PSPs is $F \cdot t \approx 1.2 \cdot 10^{-4}$g for a typical mouse of 20 g [222–224]. Analysis of tumor growth curves was conducted (Figure 4.9), and growth rates were estimated using linear approximations. In the case of control tumors, pure cell proliferation in the absence of any drug-induced cell death resulted in an average growth rate of $\Lambda \approx 70$ mm³/day, which was seen to decrease 50% in 1.0 μm PSP drug delivery to ≈ 35 mm³/day (corresponding to roughly 50% cell death, as this value is the net of cell growth minus cell death). Thus, the death rate in tumor cells due to treatment with 1.0 μm PSP delivered drug was seen to be $\Lambda_k \approx 35$ mm³/day. The specific cell kill rate (per drug molecule) is

$$\lambda_k = \Lambda_k/(Ft) \text{ and, combined with Equation 4.19, gives an estimate for } t_{kill} = \frac{2V_0}{\Lambda_k} \cdot f_{kill} \approx 4 \text{ days}$$

(where $f_{kill} \approx 0.5$ and $V_0 = 130$ mm³), which is seen to be in excellent agreement with a cell kill plateau after three days, as reported in the experiments (Figure 4.8). We note that the subsequent treatments implemented in the experimental protocol are irrelevant here, as tumors have increased in volume by one full order of magnitude by the second treatment application (i.e., one week later), which is given as the same dosage as the first treatment.

As an experimental demonstration of this concept of drug-loaded nanocarriers, we used a PSP delivery system loaded with chemotherapy in a mouse model of breast cancer. We found that these particles localize within tumors [225], likely due to the geometry of the particles. As predicted by our model, we measured a 3.5-fold difference in tumor growth rate in the mouse tumors treated with drug-loaded nanoparticles compared with the tumors treated with systemic delivery of DOX (Figure 4.8). This experimental confirmation of the model results provides rationale to translate these findings and concepts toward

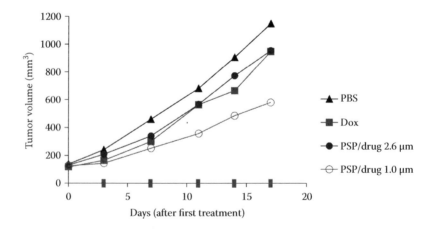

FIGURE 4.9 Measurements of tumor volume. Four treatment groups: PBS (control), free DOX, 1.0 μm porous silicon particle loaded with chemotherapy drug (PSP/drug 1.0), and 2.6 μm porous silicon particle loaded with chemotherapy drug (PSP/drug 2.6). Data were measured on days 0, 3, 7, 11, 14, and 17 after first treatment. (Reproduced from Wang, Z. et al., *PLoS Comput. Biol.*, 12(6), e1004969, 2016.)

clinical applications. Notably, we find that the pathological response to chemotherapy is heterogeneous within a given tumor, and that the local physical properties of the tumor describe this response. This is significant in understanding therapeutic resistance, suggesting that the physical microenvironment naturally selects cancer cells that reside in areas with poor drug penetration.

4.4 CONCLUSIONS

After an overview of the three papers discussed in this chapter, it is clear that tissue-scale diffusion barriers contribute to drug treatment resistance and should be accounted for in the design of a treatment plan. We have also investigated how anticancer drugs are successful in monolayer experiments, but often fail to have the same effectiveness in cancer patients. The current practice of basing treatment plans off the results of these cell-scale experiments must be replaced with a broader-scope approach that accounts for the tissue scale as well. Numerous barriers to diffusion are present at this scale: blood vessel size and distribution density, microenvironment pH, hypoxia, the dense or loose packing of cells, and cell–cell interactions are all factors that could potentially affect successful drug delivery, and thus tumor response to treatment.

We have also seen how using a physical oncology approach (focusing on developing mechanistic models to predict treatment outcomes based on measurements from the histopathology of individual patients' tumors) has allowed us to predict the fraction of tumor killed (f_{kill}) *in vitro* and *in vivo*. The feasibility of the model for clinical translation is evident in the ability to predict patient-specific f_{kill} using pretreatment computed tomography scans. The stage has been set for experiments to test the impact of other barriers on tumor response to therapy, as well as how these barriers can be overcome *in vivo* and in human patients. The potential benefits of individualized treatment plans where medical experts can accurately predict tumor response to therapy before treatment are enormous, and are becoming more realistic as advances in this exciting field move closer toward clinical integration.

CHAPTER 5

Prediction of Chemotherapy Outcome in Patients

IN CHAPTER 4, we saw that the amount of cancer killed (f_{kill}) depends on the concentration of the drug that reaches the tumor and the particular effectiveness of the drug against the tumor. Once drugs are introduced into the bloodstream, they encounter diffusion barriers, which are physiological obstructions (see Chapter 2 for more details) that prevent drugs from reaching the tumor. If adequate amounts of drug reach the tumor mass, the goal of successfully delivering chemotherapy to kill enough cells to shrink or eradicate a tumor would be realized. We thus hypothesize that if drugs can be delivered to the site of the tumor in sufficient concentration, then the drugs should perform as hoped for—as is the case in monolayer experiments where diffusion barriers do not exist. In this chapter, we show the ability of the physical oncology approach to not only understand and describe drug delivery, but also predict chemotherapeutic outcomes based on physical parameters from individual patients.

5.1 INTRODUCTION

In treating cancer with chemotherapy, a critical problem faced by oncologists is the dissimilar results of tests performed in laboratory experiments on monolayers of cancer cells in petri dishes versus treatments performed in live patients or on laboratory animals. Various cancer-fighting drugs exhibit highly effective results when delivered to monolayers, but they underperform in animal models. Theoretically, for example, if the drug can reach and be exposed to the exact same cancer cell lines *in vivo* in the same concentrations and durations as in monolayer experiments, should not the amount of cell death as a result of treatment be comparable to that of the monolayer experiments? What accounts for the difference in efficacy? Diffusion barriers are our primary target in understanding why drugs prove less effective in patients; we believe accurately describing and modeling drug transport across diffusion barriers will lead to breakthroughs for chemotherapy, even on so-called "resistant" cell lines, which show far less resistance when diffusions barriers are minimized.

The physical properties of a tumor's microenvironment influence a drug's ability to penetrate and kill tumor cells. Some of these properties can be potential obstructions to drug diffusion, which increase the tumor's resistance to chemotherapy. As described in Chapter 2, these barriers can include overexpression of protein efflux pumps, cell growth cycles, acidosis, hypoxia, tissue density, high interstitial fluid pressure, and electrostatic charge. We argued that diffusion barriers may be another main cause for a tumor's ability to be drug resistant, and while all the above factors can contribute to resistance, *the drug's primary challenge is reaching the tumor microenvironment*. Whether a drug can be ultimately delivered to the tumor (past the diffusion barriers) depends primarily on the vasculature in the surrounding area and its influence in creating a static environment that prevents perfusion of blood [226–228].

Through our integrated study together with Dr. Steven Curley (Baylor College of Medicine), we have successfully demonstrated that the key parameters in accurately predicting the amount of cancer cells that will be killed in chemotherapy treatment are the blood volume fraction (BVF) and the radius of the main blood vessels (r_b) involved in delivering blood and nutrients to the tumor. This lays the groundwork for further human clinical studies, specifically in colorectal cancer (CRC) metastatic to the liver, as well as further research into the contributions of other factors in the microenvironment to drug resistance.

5.2 MECHANISTIC MODEL

Using a three-step process, we were able to ensure accurate predictions within the patient set, not only with CRC metastatic to the liver, but also on different tumor types [7]. We will show that this model is able to describe drug diffusion within both the tumor and normal tissue, the formulation of which is based on a physical characterization of the vasculature and tissue architecture gained from histopathology.

5.2.1 Model Development

Total cancer cell kill is a function of total local drug concentration. Hence, the preliminary model (a differential equation) was designed based on diffusion and its parameters:

$$\frac{1}{r}\frac{\partial}{\partial r}\left(r\frac{\partial \sigma}{\partial r}\right) - \frac{\sigma}{L^2} = 0 \tag{5.1}$$

where σ is the local concentration of drug, r is the radial coordinate scaled with the drug diffusion penetration distance $L = \sqrt{D/\lambda}$, D is the diffusivity of the drug (a constant assuming, for simplicity, that the tumor microenvironment is isotropic), and λ is the cellular uptake rate of drug (with a unit of inverse time). Applying the boundary condition at the source's radius r_b that the intravascular drug concentration is σ_0,

$$\sigma(r = r_b) = \sigma_0 \tag{5.2}$$

a solution for the concentration distribution of drug within the tumor can be obtained:

$$\frac{\sigma}{\sigma_0} = \frac{K_0(r/L)}{K_0(r_b/L)} \tag{5.3}$$

K_0 is the modified Bessel functions of the second kind of order 0 [229]. The fraction of tumor killed (f_{kill}) is thus determined by integrating to get the volume average [230] over the region of dead tumor [229] surrounding the blood vessel:

$$f_{kill} = \frac{2 \cdot \pi \cdot h}{V_{tot}} \cdot \int_{r_b}^{r_k} f_{kill}^M(\sigma(r)) \cdot r \cdot dr \tag{5.4}$$

where r_k is the thickness of the region of dead tumor, h is the length of the blood vessel, $f_{kill}^M(\sigma)$ is the fraction of tumor cells killed in a monolayer cytotoxicity experiment (which is dependent on the local concentration of drug, σ), and V_{tot} is the total volume of tumor served by the blood vessel. Assuming that blood vessels are uniformly distributed throughout the tumor, the total volume served by an individual vessel is

$$V_{tot} = \frac{V_{tumor}}{N_{vessels}} = \frac{V_{tumor}}{V_{vessels}} \cdot \frac{V_{vessels}}{N_{vessels}} \tag{5.5}$$

where $\dfrac{V_{vessels}}{V_{tumor}}$ is the BVF, and $\dfrac{V_{vessels}}{N_{vessels}}$ is the volume of a single cylindrical blood source of radius r_b. Thus, the total volume of tumor served by the source as a function of BVF and portal radius is

$$V_{tot} = \frac{\pi \cdot r_b^2 \cdot h}{BVF} \tag{5.6}$$

Substituting Equation 5.6 into Equation 5.4 gives

$$f_{kill} = \frac{2 \cdot BVF}{r_b^2} \cdot \int_{r_b}^{r_k} f_{kill}^M(\sigma(r)) \cdot r \cdot dr \tag{5.7}$$

for the predicted fraction of tumor volume killed.

Here, we approximate the fraction of tumor cells killed in a monolayer cytotoxicity assay, that is, in the absence of diffusion gradients, $f_{kill}^M(\sigma)$, with a piecewise linear function [113]:

$$f_{kill}(\sigma) = f_{kill}^M(\sigma_0) \cdot \frac{\sigma(r) - \sigma_k}{\sigma_0 - \sigma_k} \tag{5.8}$$

($f_{kill}(\sigma) = 0$ for $\sigma(r) < \sigma_k$) where $\sigma_k = \sigma(r_k)$ is the threshold drug concentration at which the tumor cells are killed in response to the drug, given as $\dfrac{\sigma_k}{\sigma_0} = \dfrac{K_0(r_k/L)}{K_0(r_b/L)}$, and $f_{kill}^M(\sigma)$ is the fraction of cells killed by a clinically relevant dosage of drug in a monolayer experiment. Substituting Equation 5.8 and the expression for the threshold drug concentration, σ_k, into Equation 5.7 gives

$$f_{kill} = \frac{2 \cdot \text{BVF} \cdot f_{kill}^M(\sigma_0)}{r_b^2} \cdot \int_{r_b}^{r_k} \frac{K_0(r/L) - K_0(r_k/L)}{K_0(r_b/L) - K_0(r_k/L)} \cdot r \cdot dr \tag{5.9}$$

Performing the integration leads to the fraction of tumor volume killed, f_{kill}:

$$f_{kill} = f_{kill}^M(\sigma_0) \cdot \text{BVF} \cdot \frac{2Lr_b \cdot K_1(r_b/L) - 2Lr_k \cdot K_1(r_k/L) - (r_k^2 - r_b^2) \cdot K_0(r_k/L)}{r_b^2 \cdot (K_0(r_b/L) - K_0(r_k/L))} \tag{5.10}$$

$$\frac{\sigma_k}{\sigma_0} = \frac{K_0(r_k/L)}{K_0(r_b/L)} \tag{5.11}$$

Hence, f_{kill} is expressed as a function of BVF; the thickness of the dead tumor region r_k; the drug source radius r_b; K_1, the modified Bessel functions of the second kind of order 1; and the fraction of cells killed in a monolayer cytotoxicity experiment $f_{kill}^M(\sigma_0)$, all of which can be directly measured to inform the model.

The fraction of cells killed in the absence of diffusion gradients can be assumed to be $f_{kill}^M(\sigma_0) = 1$. We further assumed that the amount of drug in the liver vasculature would be sufficient to kill all cells *in vitro*. Therefore, this parameter can be omitted herein. Combining Equations 5.10 and 5.11 to eliminate σ_k and r_k, we obtain the (maximum) predicted f_{kill} for each patient as a function solely of parameters, r_b, BVF, and diffusion penetration length L, all of which can be directly measured from histopathology or imaging:

$$f_{kill} = 2 \cdot \text{BVF} \cdot L \cdot \frac{\sqrt{\text{BVF}} \cdot K_1(r_b/L) - K_1(r_b/(L \cdot \sqrt{\text{BVF}}))}{\sqrt{\text{BVF}} \cdot r_b \cdot K_0(r_b/L) \cdot (1 - \text{BVF})} \tag{5.12}$$

The model's output, f_{kill}, is dependent on the predicted concentration of drug (which is calculated by a diffusion equation) and a known effectiveness of drug at that concentration (taken from monolayer experiments). f_{kill} is calculated by averaging the drug's effectiveness over the volume of tissue around the vessel (the source of the drug). In theory, the model predicting f_{kill} (the fraction of cells killed by chemotherapy) is basic in its ideology. For the drug to kill the tumor cells, it must reach the tumor. For the drug to reach the tumor, it must travel through the surrounding vasculature and pass through the vessel walls into the tumor area. The greater the vasculature volume (the larger or more numerous the blood vessels in the tumor area), the greater the likelihood of therapeutically sufficiently high concentrations of drug reaching the tumor. Moreover, it can be concluded that the amount and size of blood vessels in a tumorous region will significantly impact the effectiveness of chemotherapy and the likelihood of diffusion barrier–induced drug resistance.

We use a three-step process to develop our mathematical model, obtain parameter values, apply the model to different tumor types, and investigate the clinical relevance in prospective studies. Step 1 of the process required designing and fitting the model to actual histopathological measurements. Once this was done and the critical parameters for prediction were found, step 2 was to validate the model on a different tumor type. The model was fit to parameters and could give accurate predictions for CRC, but the model still required experimental validation for other applications; thus, we sought to answer the question, would it have the same predictive power for other types of cancer? To test this, we used hemotoxylin and eosin (H&E)*–stained microscopic slides of glioblastoma multiforme (GBM), which is a form of brain cancer that is usually highly malignant and characterized by a plentiful blood supply. We measured BVF, r_b, and f_{kill} in these slides and input the values into the same model. We then compared the model's predicted f_{kill} with the actual measured f_{kill}, the results of which were extremely accurate, illustrating how the model has predictive power for multiple tumor types. Finally, in step 3, in order to test the predictive ability of the model, we obtained archived pretreatment computed tomography (CT) scans of CRC metastatic to the liver in 11 patients from MD Anderson Cancer Center in Houston. We then measured BVF from the CT images and accurately predicted the f_{kill} using the model. Successful validation of the model for application to other tumor types and transition toward clinical applications required the model's predicted f_{kill} to be comparable with the actual measured kill ratio from the histopathology after chemotherapy treatment.

5.2.2 Parameter Values and Model Fitting

The three main variables in the model, that is, BVF, vessel radius (r_b), and diffusion penetration distance (L), characterize the drug's ability to reach the tumor through the vasculature in the region. Measuring patient tissue from histology scans conducted before treatment provides these values that drive the output of the equation and predict f_{kill}. This ability to characterize a tumor's blood supply is essential, because successful drug delivery

---

* H&E slides are composed of tissues stained with H&E dyes. These dyes stain acidic structures a purplish color and basic structures pink. This allows tissue structures to be easily differentiated.

is dependent on blood supply to and blood sources in the tumor tissue. The more accurately an individual tumor can be physically characterized, the better the prediction of f_{kill}, thus improving the ability to create individualized treatment plans.

Since both tumor and histology sectioning are assumed to be isotropic, the fraction of tumor killed can be directly measured as the fraction of tumor *area* killed from the histopathology images. Figure 5.1A shows an illustration of the measurements of the thickness of the area occupied by dead tumor (the "kill radius" r_k in dark gray lines), from a representative histopathology specimen. These measurements were manually performed using GNU Image Manipulation Program (GIMP) [231]. To calculate the fraction of dead tumor area f_{kill}, the dead areas of tumor were colored light gray, while the live areas of tumor were colored dark gray. The fraction of dead tumor area measured from histological images was set as

$$f_{kill} = \text{\# of light gray pixels}/(\text{\# of light gray pixels} + \text{\# of dark gray pixels}) \qquad (5.13)$$

We then performed a regression analysis to determine the values for the three parameters: r_b, L, and BVF (Figure 5.1B, inset). Note that each of these estimated values is consistent with measurements from human anatomy.

BVF was found to be the single most important factor in predicting therapeutic success. Vasculature is characterized by this BVF value; the higher the BVF, the more vasculature present, which likely means a greater ability to deliver drug to the tissue. BVF was the best predictor of f_{kill} because of its description of the drug's ability to reach the tumor cells. As we have discussed, chemotherapy drugs kill cancer cells—provided they can reach them in adequate amounts first. L can be considered the kill zone for the blood vessels. L was measured as the distance from the blood vessels to the nearest live tumor cell, representing the drug's ability to extravasate from the vessels and diffuse through the tissue. Only after treatment can the amount of tumor killed around a vessel be observed, which is why L must be measured from posttreatment scans. However, L was found not to vary much from patient to patient, at least for the case of CRC metastatic to liver.

5.2.3 Prediction and Validation with CT Data

The model's accuracy was validated by taking the results from the model based on pretreatment scans and retrospectively comparing them with the measured results after treatment. Differences in output between the model's predicted f_{kill} and posttreatment observed f_{kill} were minimal, and not statistically significant, validating the model's ability to accurately predict chemotherapeutic efficacy. Throughout the model validation process, we identified that the critical parameters for prediction of f_{kill} are BVF and r_b, two parameters that can be measured using routine pretreatment CT scans.

We obtained a larger cohort of patients with CRC metastatic to liver (MD Anderson Cancer Center) where pretreatment contrast CT scans were performed, followed by chemotherapy and surgical excision. To determine if our model could predict chemotherapeutic outcome based only on standard information available from pretreatment contrast CT imaging, we carried out the following series of steps. First, we performed an analysis on

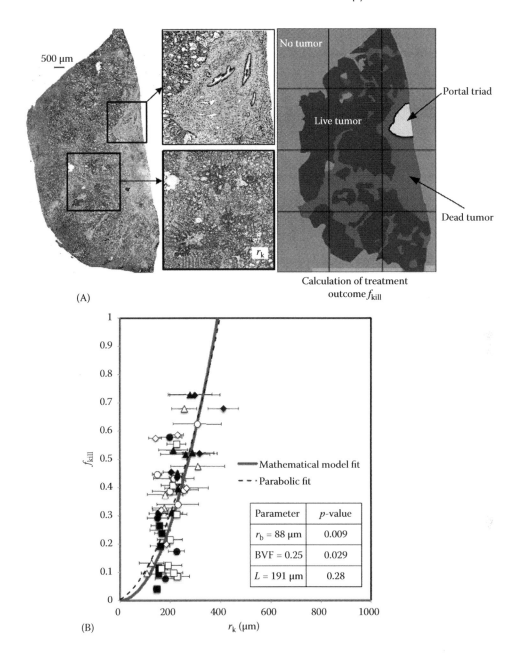

FIGURE 5.1 Mechanistic model of tumor kill from chemotherapy in patients with CRC metastatic to liver (f_{kill}). (A) Determination of model parameters from histopathological measurements. The panel on the right is a segmented, computerized image of a histologic section shown on the left. (B) The model (bold solid line) describes heterogeneous response within human cancer (symbols: 50 data points from eight patients with CRC metastatic to liver). The x-axis represents a pathological surrogate for drug diffusion penetration distance. These results demonstrate significant correlation of local tumor kill with perfusive and diffusive transport properties of tissue. Note that Equation 5.12 really depends only on r_b/L (also see [158]), meaning that L is not an independent parameter, and thus obtaining a statistically insignificant p-value for L is irrelevant here. (Reproduced from Pascal, J. et al., *Proc. Natl. Acad. Sci. U.S.A.*, 110(35), 14266–14271, 2013.)

the histopathology from posttreatment specimens similar to that described for the GBM. This again validated the predictive power of the model specifically for this third cohort of patients (Figure 5.2A, solid curve: $R^2 = 0.79$). Since we used the same value of diffusion penetration distance L and portal radius r_b obtained from the fit of the first cohort (Figure 5.1B), these results again point to uniformity of these parameters across patients, thus

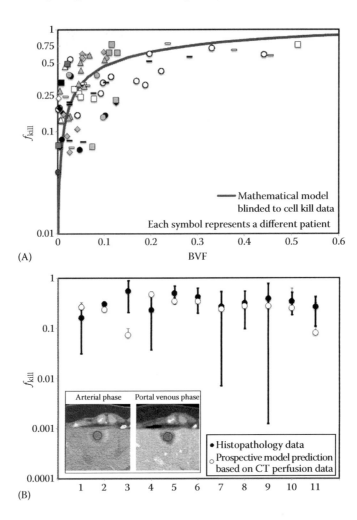

FIGURE 5.2 Prospective, patient-specific model predictions match outcomes of fraction f_{kill} of cells killed by chemotherapy in a third cohort of patients with CRC metastatic to liver. (A) Testing of Equation 5.12 from posttreatment histopathology (coefficient of determination $R^2 = 0.79$). (B) Predictions of Equation 5.12 using BVF parameter calculated (Equation 5.14) from pretreatment contrast CT perfusion measurements (open circles, average relative error ≈ 15% between prediction and actual). Multiple measurements per patient indicated by the same symbol (A) and standard deviation (B, filled circles). Model input parameters r_b (radii of liver portals) and L (drug diffusion penetration distance) from Figure 5.1B. Inset: Examples of (left) late arterial phase and (right) portal venous phase from contrast CT scans of a 2.4 cm hypodense CRC metastasis in the left anterior hepatic lobe (circles indicate example of area used for measurement). (Reproduced from Pascal, J. et al., *Proc. Natl. Acad. Sci. U.S.A.*, 110(35), 14266–14271, 2013.)

generating the hypothesis that future clinical translation would rely primarily on patient-specific determination of the parameter BVF. To test this hypothesis, we calculated a linear correlation constant for histopathology BVF and contrast CT Hounsfield units (HU), which allowed us to obtain a BVF value from the contrast enhancement of the CT images for each individual:

$$BVF = 0.00088\ CT\ (HU) \tag{5.14}$$

Inputting this value into Equation 5.12 produced accurate kill ratio predictions (Figure 5.2B, open circles) that coincided well with the actual measurements from histopathology posttreatment (Figure 5.2B, filled circles), with an average relative error of the predicted fraction killed of ≈15%. Here, the average relative error between the model prediction f_{kill} (P) and the measured kill value f_{kill} (M) was calculated as $\langle (f_{kill}(P) - f_{kill}(M))/f_{kill}(M) \rangle$.

5.3 IMPLICATIONS

This research is revolutionary in that it demonstrates how individualized treatment is not only a near possibility, but also the necessary route for the future. In theory, the model is simple: tumor death depends on drug perfusion to—and diffusion through—the tumor, but until this study, only limited information was available on why drug resistance in human tumors is so prevalent when drugs effectively kill the same tumor cells in the lab. Through physical characterization of the tumor—measuring vasculature (blood vessel prevalence and radius) and the diffusion length of the drug from the vessels—medical experts can design chemotherapy cycles to maximize the amount of tumor killed based on the diffusion of the drug in the tumor. This research can aid clinicians in accurately choosing drug dosage, dosing frequency, and type of delivery to maximize tumor kill and minimize unintentional nontarget cell death in specific patients based on their individual tumor. This model can be applied to multiple tumor types, and can be used for immunotherapy, other types of chemotherapy, and in other cases where drug delivery relies on the diffusion properties of the microenvironment.

The applicability of this model is also seemingly broad and practical, because the equations are based on parameters that can be readily measured from routine clinical tests. By only measuring three parameters, we were able to accurately predict tumor kill fractions with 85% accuracy. There are also many other parameters within a microenvironment (hypoxia, cell cycles, fibrosis, etc.) that can be further accounted for in a future model, which could aid in even more accurate predictions. Clinicians, if implementing this model alone, could tell their patients with 85% accuracy—far better than any current predictions of chemotherapy's effectiveness—how much tumor the prescribed therapy would kill. This would provide more accurate information for patients to aid in decision making, and for the clinician to be able to create the best treatment plan.

5.4 APPLICATION TO LYMPHOMA

Finally, we set out to create a platform to generate and test hypotheses of the contributions of tissue-scale parameters on chemotherapeutic response: in the process, proving that a mathematical model can predict treatment effectiveness *in vivo* based on measurements from pretreatment scans and *in vitro* experiments [158]. Using the previously described model, which is based on histopathological measurements of solid tumor characteristics, we applied the predictive ability to mice with both drug-sensitive and -resistant lymphoma cell lines.

Using the same parameters of BVF, diffusion penetration length, blood vessel radius, and fraction of cells killed *in vitro*, we attempted to predict the amount of tumor killed *in vivo*. The predictions were compared with results from live mice with either drug-sensitive or -resistant lines of non-Hodgkin's lymphoma. The cancer cells were injected into the mice and allowed to grow for 19 days before the maximum tolerated dose of doxorubicin (DOX) for a typical mouse was injected. Two days after DOX injection, the tumors were removed, sliced, and stained for cellular markers that allow vasculature and apoptosis (cell death) to be easily observed.

The parameter values were obtained after staining through manual measurements. A regression analysis was performed to fit Equation 5.12 to the measured tumor kill f_{kill} and BVF, from which we obtained the best fits for f_{kill}^{M} and r_b/L. Figure 5.3 shows a comparison of model predictions of the fraction of tumor volume killed f_{kill} from chemotherapy with those measured from the histopathological samples of drug-sensitive and drug-resistant tumors *in vivo*, and how the f_{kill} predictions change with f_{kill}^{M} and r_b/L. The model's coefficient of determination R^2 between the observed data and model predictions using the best fits for f_{kill}^{M} and r_b/L was 0.86, and hence was considered acceptable in explaining the relationship between f_{kill}^{M} and input parameters BVF and r_b/L. However, there exists noticeable variance in some experimental measurements of BVF and dead tumor area; this variance reflects the heterogeneity in tumor physical properties [232,233], which may lead to nonuniform drug penetration and tumor response.

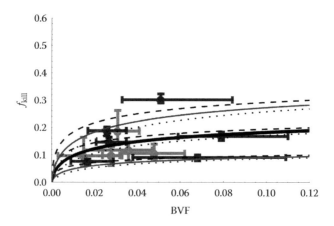

FIGURE 5.3 Comparison of model predictions with experimental measurements. The bold solid line represents the case with best fits for f_{kill}^{M} and r_b/L. Three solid lines, top to bottom: $f_{kill}^{M} = (1.5, 1.0, 0.5)$-fold of its best fit. Line scheme: $(r_b/L = (0.5, 1.0, 1.5)$-fold of its best fit) = (dashed, solid, dotted). Best fits: $f_{kill}^{M} = [0.25]$ and $r_b/L = 0.068$. (Reproduced from Frieboes, H. B. et al., *PLoS One*, 10(6), e0129433, 2015.)

We then performed a local parameter analysis (only one parameter was varied at a time, and all other parameters were held fixed at their reference values) on the model to evaluate how a change in one model parameter affected the overall system response [85,234,235]. A sensitivity coefficient was used as the evaluation index and calculated by the following equation:

$$S_p^M = \frac{(M_i - M_0)/M_0}{(p_i - p_0)/p_0} \tag{5.15}$$

where p represents the parameter that is varied, and M is the response of the system; M_0 is obtained by setting all parameters to their reference (unperturbed) values, and thus $(M_i - M_0)$ is the change in M due to the change in p, that is, $(p_i - p_0)$. In our case, the system response M corresponds to the fraction of tumor volume killed f_{kill}, and p corresponds to one of the three parameters (L, r_b, and BVF) under consideration.

For each parameter, we created 101 variations, through a range of ±50% of the parameter's reference value and with a 1.0% variation interval, according to [213]. The analysis results are shown in Figure 5.4. BVF was identified to be the most sensitive parameter, especially when low BVF values were present, followed by r_b and L. In cases where low levels of vasculature were present, even a small change in BVF led to a large change in the predicted f_{kill}, demonstrating vascularization's significant impact on the effectiveness of drug delivery.

This study demonstrated the mathematical model's ability to account for the tumor's microenvironmental effects on the efficacy of drug delivery. We found that the fraction of tumor cells killed in the center of the tumors (near the blood vessel source) was comparable between both cell lines, suggesting that if diffusion barriers can be overcome, anticancer drugs can be effective even in resistant cell lines. High tissue density in the resistant cell lines was responsible for the ineffectiveness of the treatment to the outer portions of the tumors. Drug-resistant cell lines had significantly lower responses to treatment, but we hoped to observe the effects of tissue-scale barriers on these resistant lines *in vivo*. We found that while BVF was comparable between the two cell types, the bulk of cell death in the resistant line was in the middle portions of the tumor (near a blood source), while the sensitive line showed uniform cell kill across the whole tumor. The resistant line seemed to be denser in the middle of the lymph node, creating a steeper diffusion gradient for drug delivery from the center of the lymph node (the main source of blood) to the periphery.

This study demonstrated the importance of understanding a tumor's microenvironment before treatment. BVF, the parameter that most affected the overall amount of tumor killed, could prove to be a key therapeutic target, to be optimized in order to enable drug resistance to be overcome. Currently, angiogenesis inhibitors are often used to prevent the tumor from forming new blood vessels (and thus prevent metastasis through the new vessels). Inhibiting vascularization lowers the BVF, possibly contributing to drug resistance due to lowered drug delivery. Instead, BVF could be increased during treatment, which would, according to the model, significantly increase the amount of tumor killed.

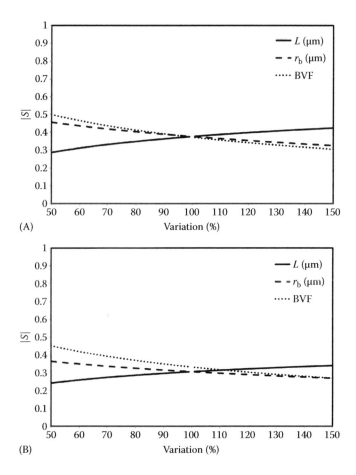

FIGURE 5.4 Model parameter perturbation analysis. Plots of sensitivity coefficients for the three parameters for (A) the drug-sensitive cell line, *Eμ-myc Arf–/–* and (B) the drug-resistant cell line, *Eμ-myc p53–/–*. S represents sensitivity coefficient. (Reproduced from Frieboes, H. B. et al., *PLoS One*, 10(6), e0129433, 2015.)

Using the same measurable parameters and mathematical model, we successfully predicted f_{kill} in mice with both drug-sensitive and -resistant lines of non-Hodgkin's lymphoma. Yet again, the model was able to successfully predict treatment outcomes for multiple types of cancer, suggesting the feasibility of predictive and individualized cancer therapy. The key to predicting treatment outcomes (the measureable parameters) could be obtained via biopsy, and BVF could be measured using pretreatment imaging. This would allow experts to accurately predict f_{kill} and thus create a treatment plan in which the drug dosing frequency to achieve maximum tumor cell death could be calculated.

5.5 CONCLUSIONS

Imagine if physicians were able to inform their patients with a high degree of confidence of these predictions: the amount of tumor a certain drug regimen would kill, the likelihood of recurrence of the cancer, and the expected length of treatment or recovery time.

Currently, drug resistance presents a major obstacle in winning the battle against most types of cancer. Physicians predict which drugs, concentrations, and delivery times will work best based on patient statistics, animal experiments, or cell culture experiments. The physicians then empirically analyze the effectiveness of the treatment and adjust the therapy depending on the cancer response and drug toxicities. Patients must wait to see how effective the treatment is, not knowing if they are being subject to unnecessary side effects from excessive dosages or are actually enabling a drug-resistant tumor to form due to clinically ineffectively low doses.

Every tumor in every patient is different. If tumors are themselves unique, should not treatment also be? One size does not fit all when it comes to cancer treatment. Individualized treatment offers hope of greater effectiveness and the ability to prevent and circumvent drug resistance, and to decrease side effects. Predictability is the key. Using the models discussed in this chapter as the groundwork for future research, clinicians will be able to base treatment plans off patient-specific parameters from individual tumors. With a high degree of confidence, they will be able to predict treatment effectiveness without ordering extra tests or any extra wait time before treatment initiation.

CHAPTER 6

Clinical Management of Pancreatic Cancer

With Jason Fleming

IN THE NEXT THREE CHAPTERS, we further demonstrate the integrated physical oncology approach through applications to pancreatic ductal adenocarcinoma (PDAC). We first discuss the clinical management and outcome of PDAC. We then show how we identify imaging-based physical markers as surrogates for molecular biomarkers, and develop mathematical models based on those physical markers for predicting treatment outcome.

6.1 INTRODUCTION

PDAC continues to be a highly lethal tumor with a very low five-year survival rate [236]. The majority of PDAC patients (~80%) are not candidates for surgical resection, which is the only potentially curative treatment. Whether a patient is a candidate for surgery or not, it is highly likely that he or she will receive chemotherapy to treat the disease at some point. However, results with drugs have been unsatisfactory, with few chemotherapy agents and limited targeted drugs showing any durable efficacy. Further, for the 10%–20% of patients with resectable disease, the results are also poor. New clinical and scientific approaches are needed for all stages of PDAC.

Patients who have metastatic disease at diagnosis receive chemotherapy, and their median survival is less than one year. Patients who do not have metastasis at diagnosis are generally classified (Table 6.1) according to whether the disease is resectable, borderline resectable, or unresectable (also called locally advanced). Those with unresectable disease generally receive chemotherapy or chemoradiation. For patients with newly diagnosed resectable and borderline resectable PDAC, the standard clinical practice is surgical resection followed by chemotherapy, with selective use of chemoradiation. The general principle with this approach is to provide control throughout the body over a disease that is potentially widespread in all patients at any given stage. However, many patients who undergo surgery for early-stage disease develop metastatic disease within a year of resection and

TABLE 6.1 Clinical Staging of Pancreatic Cancer

Localization	AHPBA/SSO/SSAT Classification [238]			MD Anderson Classification [239,240]		
	Potentially Resectable	Borderline Resectable	Locally Advanced	Potentially Resectable	Borderline Resectable	Locally Advanced
SMV/PV	No abutment or encasement	Abutment, encasement, or occlusion	Not reconstructable	Abutment or encasement without occlusion	Short-segment occlusion	Not reconstructable
SMA	No abutment or encasement	Abutment	Encasement	No abutment or encasement	Abutment	Encasement
CHA	No abutment or encasement	Abutment or short-segment encasement	Long-segment encasement	No abutment or encasement	Abutment or short-segment encasement	Long-segment encasement
Celiac trunk	No abutment or encasement	No abutment or encasement	Abutment	No abutment or encasement	Abutment	Encasement

Note: CHA: common hepatic artery; PV: portal vein; SMA: superior mesenteric artery; SMV: superior mesenteric vein.

despite chemotherapy. As an alternative to subjecting all these patients to a morbid surgical procedure, Evans and associates pioneered the idea that systemic therapy prior to resection could help select the patients with early-stage tumors who would benefit from surgery [237].

The goals of neoadjuvant therapy are to increase the probability of successful surgery and to reduce the risk of local and distant recurrence. Two decades of research have shown this approach to be safe and well tolerated [241]. In addition to identifying patients who have aggressive biology and who would not have benefited from a radical surgery, the preoperative therapy provides some prognostic information for those who do undergo resection, as the extent of pathological response to therapy has been associated with outcomes [242]. However, only a small minority of patients achieve an excellent response to neoadjuvant therapy (less than 10% viable tumor cells), and methods to identify these patients *a priori* are currently lacking in the clinic. Namely, with the exception of CA19-9 (see "Definition of Technical Terms"), a biomarker with several limitations, there are no viable prognostic or predictive biomarkers for PDAC [243].

We have attempted to address the lack of useful biomarkers for PDAC by focusing on the physical properties of PDAC [8]. This approach may help in the design of rational therapeutic strategies to improve responses for this aggressive disease. Moreover, recent studies suggest that this approach will have application to emerging therapeutic strategies. In this chapter, we review previous attempts at targeted therapies for PDAC and explain how physical sciences–based approaches may improve our understanding of PDAC and the clinical management of this disease.

6.2 MULTISCALE TRANSPORT DYSREGULATION IN PDAC

PDAC has significantly different physical properties than the surrounding pancreas parenchyma (Figure 6.1), and these properties differ from patient to patient. For drugs and nutrients to reach the cells within PDAC, they must traverse dysfunctional blood vessels and extensive amounts of stroma that surround the cancer cells. Some drugs and molecules must pass through cellular transporters, which can have variable expression. The process of moving through each of these physical aberrations in the tumor is a prime example of a multiscale transport phenomenon and has been called the idea of transport oncophysics (see "Definition of Technical Terms"), which views cancer as a series of physical processes that have become unhinged [8].

In Chapter 7, we describe the full details of our clinical trial of intraoperative gemcitabine infusion during resection of PDAC. This trial was the first of its kind and supported the oncophysics view of cancer. Specifically, we showed that gemcitabine DNA incorporation into tumor cells depended on the microenvironment (i.e., stroma; see "Definition of Technical Terms") and cellular factors (i.e., human equilibrative nucleoside transporter [hENT1]; see "Definition of Technical Terms") [244]. This result may explain the failure of some clinical trials to show a benefit of novel gemcitabine formulations. Namely, these trials did not take into account the multiscale transport dysregulation of PDAC. For instance, a gemcitabine drug–lipid conjugate did not show a survival difference in patients with low hENT1 expression compared with gemcitabine alone [245], possibly because of the lack of

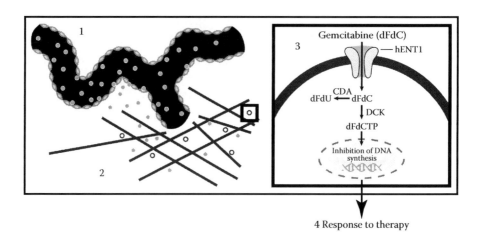

FIGURE 6.1 For gemcitabine (circles inside blood vessels) to reach pancreatic cancer cells (open circles outside blood vessels), it must go through (1) the deranged and sparse blood vessels of the tumor, (2) the ECM and stroma that surround the tumor cells, and (3) the membrane transporter hENT1. Inside the cell, gemcitabine (dFdC) undergoes enzymatic processing and then reaches the DNA of the cell, which results in a (4) response to therapy. Cytidine deaminase (CDA); deoxycytidine kinase (DCK). (Reproduced with permission from Koay, E. J. et al., *J. Clin. Invest.*, 124(4), 1525–1536, 2014.)

stratification according to other transport characteristics, especially the variable stromal content of the tumors.

Through the identification of physical characteristics that have scientific and clinical significance, the concepts of physical oncology, such as oncophysics, could have a major impact on PDAC. For example, the characterization of the physical properties of PDAC can be considered physical biomarkers, and these may lead to more effective and personalized PDAC treatment. In addition to cytotoxic agents like gemcitabine, this approach has application to numerous therapeutic strategies for PDAC, including inhibition of angiogenesis, stromal modification, and metabolic therapy. In the subsequent sections, we will discuss these treatment strategies in the context of physical stratification methods.

6.3 INHIBITION OF VEGF IN PDAC

About two decades ago, there was significant interest in inhibiting vascular endothelial growth factor (VEGF; see "Definition of Technical Terms") in PDAC because VEGF expression was thought to protect the tumor endothelium and tumor cells from cytotoxic agents and radiation [246]. It was well recognized that angiogenesis is a fundamental phenomenon associated with the development and progression of almost every type of cancer, including PDAC [247–250], even though PDAC is generally considered a hypovascular tumor. Neovascularization promotes the growth of tumor cells by nourishing them with oxygen and nutrients. The process usually begins when transformed cells begin secreting stromal-modifying proteins, including angiogenic factors such as VEGF, and later may involve synergistic cooperation between tumor and stromal cells [251].

TABLE 6.2   Clinical Trials of Bevacizumab for Advanced Pancreatic Cancer

Study	Intervention	Endpoint	Result
Varadhachary et al., JCO 2006 [239]	Phase I: Bev + Cape + XRT ($n = 47$)	Toxicity and MS	No significant toxicity after excluding those with duodenal involvement, MS of 11.6 months
RTOG 0411 [252]	Phase II: Bev + Cape + XRT ($n = 82$)	1-year survival rate	47% surviving at 1 year
CALGB 80303 [253]	Phase III: Gem + Bev vs. Gem + placebo ($n = 602$)	OS	No difference in OS (5.8 vs. 5.9 months)

Note: Bev, bevacizumab; Cape, capecitabine (see "Definition of Technical Terms"); Gem, gemcitabine; JCO, *Journal of Clinical Oncology*; MS, median survival; OS, overall survival; XRT, radiation.

A number of trials attempted combining bevacizumab (a monoclonal anti-VEGF antibody) with cytotoxic agents like radiation and chemotherapy. Table 6.2 summarizes trials of bevacizumab. Unfortunately, these studies were considered failures, and bevacizumab and other inhibitors of VEGF have not been pursued for PDAC with any enthusiasm since. However, these trials were not stratified according to any markers related to angiogenesis. Interestingly, there were a handful of patients in these clinical trials who survived beyond two years, as seen on the survival curves. This exceeds the typical survival time for patients with advanced PDAC, and suggests that the therapy may have been active in these patients [252,253]. The biological and physical differences of these tumors may be anticipated to be relatively more vascular and dependent on VEGF for tumor metastasis. Thus, stratification and selection of patients according to these properties may enable a revival of this strategy for PDAC.

6.4 STROMAL MODIFICATION

PDAC is commonly associated with an extensive degree of desmoplasia, which is the growth of fibrous tissue. This is also called a stromal reaction. The variability of stromal composition within a single tumor in the pancreas and between patients with PDAC suggests a diverse combination and proportion of stromal components: fibroblasts, immune cells, and endothelial cells embedded within an extracellular matrix (ECM) [254]. In this section, we analyze some of the major components and pathways related to the ECM that are currently under investigation for PDAC and explain how these may be quantified in terms of their physical properties, and these properties may in turn be useful for the selection of patients for therapeutic strategies that target the stroma.

6.4.1 Complex Role of the Stroma in PDAC

For decades, it has been known that the stroma supports pancreatic cancer progression, but many of these studies were conducted *in vitro* [255]. Moreover, it was generally assumed that the stroma also acted as a physical barrier to drug delivery. Genetically engineered animal models of PDAC have been generated to study these interactions [217,256–258], and it is clear that the role of the stroma in PDAC is complex.

For example, Sonic Hedgehog (Shh), a key pathway that promotes stromal desmoplasia, has been genetically deleted or pharmacologically inhibited to reduce desmoplasia in

PDAC. This intervention led to an accelerated tumor growth, increased systemic morbidity, increased metastasis, and decreased overall survival in animals [257]. Furthermore, tumors lacking the Hedgehog signaling pathway were smaller despite a more aggressive and lethal phenotype, a finding that may reflect that stroma normally comprises a large percentage of PDAC tumor volume [254]. The precise biological mechanisms that explain these observations have not been defined and are under investigation.

These studies highlight that stromal depletion strategies must be used with caution and with good scientific rationale. Notably, other studies that preceded the stromal investigations by Rhim et al. [257] demonstrated improved drug delivery and efficacy with the approach. The Shh signaling pathway contributes to stromal desmoplasia in multiple solid tumor systems, including PDAC. This developmental morphogen pathway is normally inactivated in normal adult pancreas but is reactivated in inflammatory and neoplastic events [257,259]. Specific inhibition of the Shh pathway depleted the tumor stroma, which in turn caused a transient increase in intratumoral perfusion and increased the delivery of gemcitabine to the tumor in a mouse model of PDAC [217], but this was not observed consistently in humans [260]. Notably, a Phase I trial of FOLFIRINOX plus IPI-926, a Shh inhibitor, showed a high objective response rate (67%) but was closed early because a separate unpublished Phase II trial of IPI-926 plus gemcitabine indicated detrimental effects of the combination [260]. Neither of these trials stratified patients according to stromal content within the tumors.

6.4.2 Targeting Specific Stromal Molecules and Extracellular Components

Connective tissue growth factor (CTGF/CCN2) has also been targeted as a way to disrupt the stromal microenvironment of PDAC because it is overexpressed in human PDAC. FG-3019 is a monoclonal antibody specific for CTGF, and its administration to KPC mice resulted in increased induction of tumor cell apoptosis when combined with gemcitabine [261,262]. This has led to Phase I/II studies in humans.

Beyond stromal-related pathways and growth factors, specific molecules have been investigated. Within the PDAC microenvironment, hyaluronic acid (HA) is abundant and disorganized, inviting speculations on its role in disease biology and resistance. HA is one of the most hygroscopic molecules in nature, and this primary characteristic of trapping water is likely the key contributor to many of its attributed biological functions [263]. As part of the desmoplastic reaction of PDAC, HA may contribute to an increased interstitial fluid pressure (IFP) within the tumor that limits its perfusion and decreases intratumoral oxygen. Enzymatic degradation of HA resulted in a rapid reduction of IFP, accompanied by the appearance of widely patent and functioning blood vessels. In mouse models of PDAC, combining enzymatic degradation of HA with gemcitabine improved survival relative to gemcitabine alone, highlighting a method to increase the sensitivity of the tumor cells to conventional cytotoxic agents [218].

This preclinical work has led to two clinical trials that are currently recruiting patients. NCT01959139 is a partially randomized Phase I/II trial comparing combined FOLFIRINOX and pegylated human hyaluronidase (PEGPH20) with FOLFIRINOX alone in patients with newly diagnosed metastatic pancreatic cancer. NCT01839487 is a Phase II trial that will compare the treatment effect of PEGPH20 combined with nab-paclitaxel and gemcitabine with that of nab-paclitaxel and gemcitabine alone in metastatic pancreatic cancer.

In addition to HA, other ECM proteins play key roles in tumorigenesis. For example, secreted protein acidic and rich in cysteine (SPARC) plays a critical role in tissue homeostasis, regulating ECM structure through interactions with albumin, collagens, matrix metalloproteinases, and growth factors. These interactions affect angiogenesis [264], transforming growth factor-β (TGF-β) signaling, and epithelial-to-mesenchymal transition and invasion [265]. Clinical and pathological studies of SPARC have suggested that its expression is a negative prognostic factor in patients with PDAC [266,267]. Because of SPARC's high affinity for albumin, this protein has gained attention in clinical trials that use nab-paclitaxel, which is an albumin–drug complex [268].

We recently identified lumican as another ECM molecule that has high expression in PDAC. Lumican is a small leucine-rich proteoglycan (SLRP) [269]. SLRPs are known to modulate cell migration and proliferation during embryonic life, tumor development, and tissue repair. Additionally, SLRPs in the ECM modulate tissue hydration and fibrillar collagen formation [270]. Lumican has been shown to modify ECM by inducing endothelial cell apoptosis, and this antiangiogenic function can explain the reduction in size of tumors expressing lumican in *in vivo* models [271]. Lumican also cross talks with cancer cells as it inhibits tumor cell glycolytic metabolism and induces cell apoptosis [272]. The expression of extracellular lumican in the primary PDAC tumors of patients was associated with significantly lower rates of distant metastasis and higher rates of overall survival [272]. Thus, lumican expression may be used as a biomarker in PDAC.

The stroma in PDAC is an area of intense interest because various pathways (e.g., Shh) and stromal components (e.g., SPARC and lumican) influence the processes of invasion, metastasis, and therapeutic response. Efforts to modulate the stroma through targeted therapies may increase drug delivery to tumor cells, but may also deplete a biological component of the tumor that is restricting metastatic spread. Rational treatment strategies will need to be designed with these different roles of the stroma in mind. Further, the heterogeneity in stromal amount and composition should also be considered in these strategies, as some patients will be less likely to benefit from such approaches. In this regard, characterization of the physical properties of the PDAC tumors may enable a stratification strategy, especially since many of the stromal components influence angiogenesis and collagen production and assembly, which are properties that could be measured with a variety of diagnostic imaging modalities, including contrast-enhanced computed tomography (CT) scans, magnetic resonance imaging (MRI), and positron emission tomography (PET).

6.5 TUMOR METABOLISM

For decades, it has been suspected that abnormal glucose metabolism may play a role in the etiology of pancreatic cancer. A seminal study demonstrated the link between abnormal postload glucose levels and subsequent development of pancreatic cancer in thousands of subjects who were followed for decades [273]. More recent mechanistic studies have shown that PDAC cells have an increased dependency on the amino acid glutamine to fuel anabolic processes, which are required for tumor growth. Furthermore, this dependency on glutamine stems from reprogramming of glutamine metabolism that is mediated by oncogenic KRAS through transcriptional upregulation and repression of key metabolic enzymes [274]. This altered metabolism results in a marked increase in the NADPH/NADP+ ratio, likely helping to maintain the cellular redox state.

This glycolytic-dominant metabolism yields high levels of lactic acid that is secreted in the tumor microenvironment (TME). This acidotic environment favors tumor invasion and metastasis. It also decreases drug efficacy. Furthermore, the acidotic environment suppresses anticancer immune cell activity with decreased cytotoxic T-cell cytokine production [275,276]. T cells rely on glycolysis as well, and they pump lactic acid into the environment; however, due to high lactate levels in the microenvironment and the unfavorable concentration gradient, lactate accumulates in T cells. Subsequently, T cells are self-inactivated by their own production of lactic acid from glycolysis. This demonstrates how cancer cells modulate the TME and tumor-reactive T cells in a tumor growth–favoring manner [276]. Novel therapies have been designed to target this metabolic pathway. For instance, inhibition of glucose transport in gemcitabine-resistant PDAC inhibited tumor growth and restored gemcitabine sensitivity *in vivo* [277].

While glycolysis appears to play a major role in PDAC, oxidative phosphorylation has also been revealed to be a critical factor [258]. In particular, a subpopulation of dormant tumor cells after oncogenic KRAS ablation had features of cancer stem cells and relied on oxidative phosphorylation for survival. Through inhibition of oxidative phosphorylation, tumor recurrence in a mouse model was markedly decreased, suggesting a role for metabolic targeting of PDAC [258]. It is currently unclear how this strategy should be employed in specific patient populations. One may hypothesize that the patients who have early recurrence after surgical resection would mostly likely benefit from such a strategy. Recurrent disease can occur in 30%–50% of patients within the first six months. Identifying these patients *a priori* has been challenging, however. Given the heterogeneity of PDAC, selection of patients based on the metabolic dependencies of their tumors may be necessary for the clinical evaluation of therapies that target the energy production of PDAC.

6.6 IMMUNOTHERAPY

Thus far, single-agent immune-based therapy has not shown significant responses in patients with PDAC [278,279]. These unfortunate results are in spite of the fact that almost all patients with PDAC have CD4+ and CD8+ T cells in their bone marrow that are reactive against PDAC antigens, and about half of the patients have circulating and tumor-invading reactive CD4+ and CD8+ cells [280]. For cancer to survive the host immune system and immunotherapy, it develops mechanisms to evade immune surveillance. These mechanisms include increased T-cell apoptosis; upregulation of immune suppressors such as interleukin-10 (IL-10), TGF-β, and VEGF; immune checkpoints; and recruitment of immunosuppressive CD4+CD25+FoxP3+ T regulatory cells, creating a favorable environment for more unrestricted tumor growth and progression [281,282]. Covering these escape mechanisms is essential for effective immune-based therapeutics.

To a certain degree, the TME forms a barrier to tumor-reactive immune cell infiltration, and it constitutes an unfavorable environment for immune cell activity and survival

[283], partly due to the high degree of lactate in the TME. Although PDAC is generally regarded as an immunosuppressive tumor, the degree of cytotoxic T-cell infiltration inside the tumor has been associated with better patient survival, with preferential distribution along the invasive border of the tumor rather than the tumor center [284]. This immune reaction plays a principal role in modulating the tumor microenvironment [284]. Undoubtedly, it is necessary to develop immunotherapy drugs that target tumor cell antigens, enhance tumor immunogenicity, or modulate the TME to maximize the antitumor activity of immunotherapy [283].

Programmed death-1 (PD-1) and programmed death ligand-1 (PD-L1) serve as key regulators of the immune response, called checkpoints. Activated T-cells express PD-1, while cancer cells and the associated stroma express PD-L1 and PD-L2. Upon binding of PD-1 to one of these ligands, T cells inactivate and undergo apoptosis, allowing tumor immune escape [285,286]. Clinical trials with monoclonal antibodies targeting PD-L1 or PD-1 have shown promising results in several tumors, such as renal cell carcinoma, melanoma, and non-small-cell lung cancer [287]. Although PD-L1 expression has been associated with poor prognosis in PDAC, targeting PD-L1 in PDAC has not shown significant response [286].

The poor response of PDAC to checkpoint blockade may be due to factors that modulate this immune checkpoint molecule. Cancer-associated fibroblasts (CAFs) secrete high levels of chemokine (C-X-C motif) ligand 12 (CXCL12), a ligand for C-X-C motif chemokine receptor 4 (CXCR4) in mouse models. Expression of this chemokine promotes an immunosuppressive microenvironment. However, targeting TME components such as fibroblast activation protein (FAP)-expressing CAFs synergizes anti-PD-L1 activity and has resulted in increased CD3 T cells infiltrating the tumor and tumor regression [288]. The PD-1-PD-L1 coinhibitory axis has been targeted in recent preclinical studies. Transduction of T cells with PD-1-CD28 fusion receptor reverted the inhibitory PD-1-PD-L1 axis into CD28 costimulation, resulting in a decrease in peripheral anergy (tolerance) and increased cytokine production, T-cell proliferation, and tumor cell lysis [289]. Indeed, the presence of cytokines secreted by activated and functioning T cells may affect TME vascularity and fibroblast recruitment. Nomi et al. found that the addition of a PD-L1 blockade increased the chemosensitivity of PDAC to gemcitabine in mouse models [286]. Partial restoration of tumor angiogenesis and stroma after the recovery of tumor-reactive T-cell function may increase gemcitabine delivery and explain, in part, this observed synergism [290]. Presently, neoadjuvant anti-PD-1 antibody (CT-011) combined with gemcitabine is being investigated in a clinical trial (NCT01313416).

Vaccination-based immunotherapy has been investigated as well. For example, combined dendritic cell–based vaccination and gemcitabine showed increased survival in a murine PDAC model [291]. Lutz et al. attempted to convert PDAC from nonimmunogenic to immunogenic tumor using GVAX, a vaccine composed of two irradiated granulocyte-macrophage colony-stimulating factor–secreting allogeneic PDAC cell lines. After neoadjuvant immunotherapy in patients with resectable PDAC, tumors harbored tertiary lymphoid aggregates resembling lymph nodes. Notably, the vaccination induced a high intratumoral $T_{effector}/T_{reg}$ ratio [283]. Recently, Le et al. showed a survival benefit of a heterologous prime or boost with cyclophosphamide/GVAX and CRS-207 over cyclophosphamide/GVAX alone. CRS-207 is a recombinant live-attenuated double-deleted *Listeria monocytogenes*, which was engineered to secrete mesothelin into the cytosol of infected antigen presentation cells [292]. Although the combinatorial strategy was able to improve survival, the effect was not durable, as the median overall survival was 6.1 months.

Another strategy to modulate the effects of the immune system and cytotoxic therapies involves chemo-immuno-radiation in pancreatic cancer (NCT01903083). This trial is based on the observation that high-dose-per-fraction radiation is associated with robust antitumor immune response, but also can be associated with a strong suppressive myeloid response. The immune modulator tadalafil has been shown to reduce myeloid-derived suppressive cell (MDSC) function and can target the suppressive myeloid response associated with hypofractionated radiation. Patients with locally advanced or borderline resectable pancreatic cancer are eligible. The patients will be treated with tadalafil for the duration of the trial. The primary endpoint of this study is to evaluate the safety of this combination treatment.

In the evaluation of these immunotherapies, the stratification and selection of patients for different strategies will be vital. Recent data have shown that patients with mismatch repair deficiencies have a higher somatic mutation load, higher response rate, and prolonged progression-free survival after immunotherapy compared with those without mismatch repair deficiencies [293]. This supports the use of different immunotherapy strategies based on a biological property of the tumors. It is notable that the responses in this trial varied considerably. Thus, modulation of the immune response by other factors should be considered. In this regard, the baseline immune cell infiltration into the tumor, the vascular properties of the cancer, and the presence of other immune-modulating stromal elements may factor into the equation. Characterization of these properties may enable rational approaches to immunotherapy in patients.

6.7 CONCLUSIONS

Prior attempts to translate promising preclinical results with targeted agents to patients with PDAC have likely failed due to the lack of patient selection, including erlotinib [294], bevacizumab [253], and Shh inhibitors [295]. The heterogeneity between patients and within tumors (Figure 6.1) is substantial. This cautions against a one-size-fits-all therapeutic approach to PDAC. Such an approach would be unlikely to show substantial survival differences. One novel approach to stratifying patients involves the characterization of the physical properties of PDAC, which are intimately tied to its biology. The vascular, stromal, immune, and metabolic properties of PDAC all relate to or are influenced by physical properties of the tumor. The development of quantitative, physical biomarkers of PDAC may complement biological approaches and help in the design and interpretation of clinical trials that aim to improve the outcomes for this deadly disease.

CHAPTER 7

Application of Cancer Physics in the Clinic

A S WE HAVE SHOWN, cancer treatment is often ineffective due to factors such as the inability to accurately predict outcomes to chemotherapy and the presence of diffusion barriers at the tissue scale. Using a mathematical pathology approach would allow physicians to accurately characterize a tumor, better understanding its microenvironmental properties. As a result, physicians would be better equipped to create effective treatment plans. Thus far, the research we have covered has consisted of *in vitro* experiments, *in vivo* experiments in mice, and patient data analysis to some level, but here we discuss our most recent breakthroughs, which include *a first-of-its-kind clinical trial to demonstrate our concepts in human patients* [8]. We describe how studying the mass transport properties of tumors can give insights into the response to chemotherapy treatment of pancreatic cancer.

7.1 PANCREATIC CANCER

In 2010, pancreatic cancer was the fourth leading killer among cancers in the United States, accounting for 7% of all cancer deaths (approximately 40,000 people this year) [296]. By 2030, it is projected to be the second most deadly cancer [297]. Despite two decades of research, treatment outcomes have not improved much; the five-year survival rate is only 6.7% [209]. Causes for these poor outcomes include difficulty in early detection, the ability of the cancer to spread rapidly to other organs in the body, a high rate of recurrence, and resistance to current therapies. While these characteristics contribute to this cancer's ability to survive and spread, the reasons for the poor outcomes are not uniform. For example, significant variations in responses to therapy have been observed between different patients and even within individual tumors [242,298–300]. Preclinical and clinic studies suggest that this cancer's ability to resist therapy may be partly attributed to ineffective chemotherapy delivery to the cancer cells [217,301].

Chemotherapy drugs for pancreatic cancer are usually given intravenously and must pass a series of physical barriers as the drug travels from the bloodstream to the site of the tumor and disrupts intracellular processes, such as DNA replication or microtubule formation. Pancreatic ductal adenocarcinoma (PDAC), the most common type of pancreatic cancer, is characterized by a leaky and disorganized vasculature, abnormally dense stromal cells, and deregulated transport proteins [218,302,303]. These features are physical barriers to the drug delivery process. En route to an intracellular target, the drug must first diffuse across blood vessels into the surrounding tissue. Here, the leaky vasculature causes excess pressure in the interstitium—the fluid region between cells—preventing the drug from reaching the extracellular region. Drug transport is further inhibited in this region due to the presence of dense stroma (i.e., collagen and other extracellular matrix) that comprises much of the tumor. Once past the extracellular region, the drug must cross the cell membrane. Usually, protein transporters provide a pathway, but deregulation of these proteins in PDAC tumors can prevent the drug from entering the cell. These physical characteristics of the tumor's microenvironment contribute to its *mass transport properties*— the capacity of a medium to permit the flow or motion of something through it. The mass transport properties of tumors tend to be much different than those of normal tissue; thus, studying a tumor's physical properties may provide a deeper understanding of resistance to drug delivery and the poor outcomes of PDAC.

One of the ways to measure mass transport properties is through diagnostic imaging. If a patient is suspected of having pancreatic cancer, a doctor will order an imaging test to get a clear visual of the organ. Computed tomography (CT) scans are a common form of imaging test where a series of x-rays are processed to form a cross-sectional or even three-dimensional image of the pancreas. These images aid in diagnosis and treatment planning; only certain PDAC tumors are removable, and the course of treatment depends on the stage, size, and spread of the tumor. The pancreatic protocol CT scan utilizes iodine-based contrasts given intravenously, as well as time-phased scans to provide a clear image of the different structures. These phases begin before the contrast is given and at specific time points from seconds to minutes after injection. Earlier phases tend to show the contrast still in the arteries and in less dense regions, while in later phases, denser areas are visible [304–306]. Tumors are characteristically less dense than normal pancreatic tissue after injecting contrast, allowing the size and location of the tumor to be distinguishable in most cases. Once the tumor is imaged and a biopsy taken, it can be diagnosed and staged, and a treatment plan can be developed. Here, we describe how we can extract further information about the tumor.

7.2 MODELING DRUG RESPONSE IN PDAC TREATMENT

As we have covered in previous chapters, physical oncologists and mathematicians are currently partnering to further understand drug resistance in tumors, as well as develop methods to overcome resistance. Using parameters from individual patients, these researchers work to develop models that describe drug delivery and potential barriers of delivery to tumors. Modeling includes the entire process of cancer treatment, from delivery to uptake by the tumor to the effects of the drug on a subcellular level. The end goal of the work being done in this field is to provide patient-specific treatment for cancer with better outcomes and fewer side effects than traditional chemotherapy. For mathematical modeling to be clinically relevant, it must be able to be easily implemented in the diagnosis and treatment process. If developing a model of a specific tumor requires many extra steps and is inconvenient for physicians and patients, it is unlikely to be implemented. Encouragingly, we have developed a model and analysis technique that uses standard-of-care imaging for patients with PDAC.

7.2.1 Mass Transport Model

We anticipate that mathematical modeling of the changes in enhancement of the tissues (measured in Hounsfield units [HU]) at sequential time points during the pancreatic protocol (precontrast, arterial, and portal venous phases) could quantify the mass transport properties of individual PDACs. We used multiple, systematic measurements obtained during the pancreatic protocol CT in a novel mathematical model to yield phenomenological parameters of mass transport that describe qualities of the pancreatic tissue (normal and tumor) and its surrounding vasculature.

The transport model consists of one ordinary differential equation (ODE) describing the variable density $Y(t)$ (in HU) in the tissue as a function of time t resulting from transfer of contrast agent molecules through the vessel walls at rate R (in s^{-1}) and clearance rate from the vasculature R_c (in s^{-1}):

$$\frac{dY}{dt} = R \cdot \left(Y_{max}^{V} \cdot e^{-R_c \cdot t} - Y \right) \tag{7.1}$$

where Y_{max}^{V} represents the (imposed) level of density within the microvasculature. Equation 7.1 is solved for initial condition $Y(0) = 0$, giving the solution for $Y(t)$:

$$Y(t) = Y_{max}^{V} \cdot R \cdot \frac{e^{-R_c \cdot t} - e^{-R \cdot t}}{R - R_c} \tag{7.2}$$

Two other model parameters can be derived from the intrinsic variables of the model (R, R_c, and Y_{max}^{V}): maximum enhancement of tissue $\left(Y_{max}^{T} \right)$ and initial influx rate of contrast (R_0). The model function can also be integrated over time to give an area under the model-predicted enhancement curve (AUC).

7.2.2 CT-Derived Mass Transport Properties

To understand the drug delivery and response relationships in PDAC, we described a method to measure the mass transport properties of PDAC tumors [8]. We used measurements from 176 routine CT scans taken before treatment to create a model that can quantify the mass transport properties of normal and malignant pancreatic tissues (Figure 7.1). In this study, we were able to quantify the differences between tumor and normal tissue, which is consistent with clinical observations that PDAC tumors are hypodense (Figure 7.2). Out of the

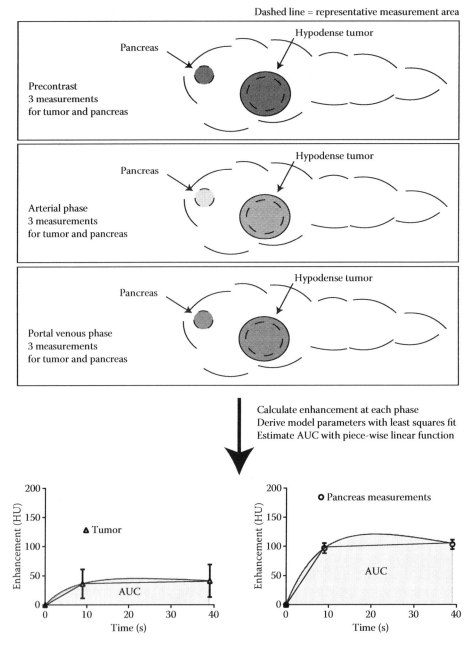

FIGURE 7.1 An illustration of the measurement technique is shown. The pancreatic protocol involves multiple phases of CT scan that are timed in relation to the injection of iodine-based contrast. The mathematical model used knowledge of this timing, as well as general knowledge about the time it takes for contrast to reach the pancreas. The model curves are shown for tumor and normal pancreas of one patient. The model can be approximated using a piecewise linear function. (Reproduced with permission from Koay, E. J. et al., *J. Clin. Invest.*, 124(4), 1525–1536, 2014.)

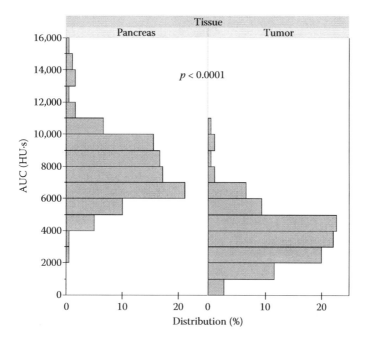

FIGURE 7.2 The distributions of mass transport properties from the CT scans of 176 patients are shown for the pancreas tissue and the tumors. The tumors exhibit worsened transport properties relative to the normal pancreas. (Reproduced with permission from Koay, E. J. et al., *J. Clin. Invest.,* 124(4), 1525–1536, 2014.)

176 scans, 110 patients received preoperative gemcitabine-based chemoradiotherapy. We were also able to predict pathological response to drug treatment and even patient survival of these 110 patients solely based on the properties derived from the pretreatment CT scans.

To assess the appropriateness of the developed model, we compared the model-derived AUC with an estimate of AUC using a simple piecewise linear function. There was a 1:1 linear relationship between the estimated and model-derived AUC parameters (Figure 7.3), which was expected since our model is continuous; moreover, the simple piecewise linear approximation can be easily translated to clinical practice, as only a straightforward calculation is necessary.

In a follow-up study [307], we found significant heterogeneity in the mass transport properties between the outer and inner regions of individual tumors. These properties of the outer and inner regions of each tumor were derived from scans done after surgical resection of PDAC tumors, finding minimal to 200% differences in mass transport properties (Figure 7.4). Within an individual tumor, the inner portion might be very susceptible to treatment, while the outer portion might be very resistant to drug transport. From a single biopsy at a focused location in the tumor, tumor characteristics at that point may not be representative of the entire tumor. Both studies showed that with the predictive results of the model, it might no longer be necessary for a patient to wait for treatment to begin to evaluate the effectiveness of the drug therapy. Instead, mass transport properties of tumors can potentially be verified prior to treatment, allowing the effectiveness of the drug to be accurately predicted. From these valuations of transport properties, the most effective plan for treatment can be implemented.

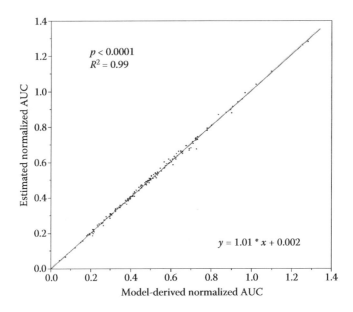

FIGURE 7.3 The estimated normalized AUC was calculated using a simple piecewise linear equation, and compared with the model-derived normalized AUC. This demonstrated a 1:1 correlation between the estimated and model-derived values, illustrating how the continuous model function (Equation 7.2) can be approximated using a straightforward calculation that can be applied at any institution. (Reproduced with permission from Koay, E. J. et al., *J. Clin. Invest.*, 124(4), 1525–1536, 2014.)

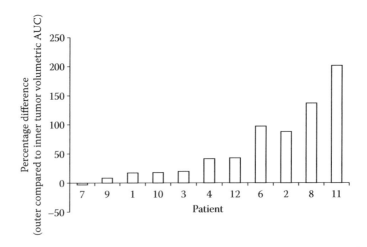

FIGURE 7.4 The intratumoral heterogeneity of mass transport properties is shown based on measurements of the normalized AUC derived from CT scans from the inner and outer regions of PDAC tumors. (Reproduced with permission from Koay, E. J. et al., *J. Clin. Invest.*, 124(4), 1525–1536, 2014.)

7.2.3 Drug Delivery and Mass Transport Properties

To prove that gemcitabine reaches tumor cells in PDAC and to determine the factors that influence drug delivery, we performed a first-of-its-kind clinical trial where 12 patients were injected with gemcitabine during PDAC resection surgery. At the beginning of the operation, the patients were intravenously injected with 1000 mg/m^2 of gemcitabine at a constant rate until the tumor was removed. Once removed, the tumors were halved by a pathologist and biopsies were taken from the outer and inner portions of the tumor. The outer portion was distinguished as the outermost 3–5 mm of tissue, while the inner portion was anything inside of that.

We sought to determine whether the variability of gemcitabine incorporation in tumors among the patients could be explained by mass transport phenomena. Because extensive desmoplasia is a common feature of PDAC that reflects a putative stromal barrier to gemcitabine delivery and also impairs vasculature function [217,218,266,303], the stromal amount in the specimens may influence gemcitabine delivery. Considering that expression of the nucleoside transporter of gemcitabine, human equilibrative nucleoside transporter (hENT1), correlates with outcome after adjuvant gemcitabine therapy in patients with PDAC [308] and our initial multiscale mass transport hypothesis [244], hENT1 scoring may improve our correlations between stromal score and normalized gemcitabine incorporation. We first ranked patients by hENT1 staining intensity and then assigned designations of high and low down the ranked order. We subsequently assessed correlations between stromal score and normalized gemcitabine incorporation for the high hENT1 and low hENT1 groups. We then identified five patients with high hENT1 staining and seven patients with low hENT1 staining. In this manner, we found that the hENT1 score significantly correlated with normalized gemcitabine incorporation (Figure 7.5A). Moreover, the stromal score inversely correlated with normalized gemcitabine incorporation after accounting for the hENT1 score (Figure 7.5B), as we had initially anticipated. We derived mass transport parameters for 11 of the 12 total clinical trial patients who had pretherapy pancreatic protocol CT scans, and noted significant correlations between the CT-derived parameters and tumor stromal score within the surgical specimen (Figure 7.5C). Moreover, the normalized CT-derived parameters were also inversely correlated with tumor gemcitabine incorporation (Figure 7.5D).

CT-derived transport parameters correlate with response to and survival after gemcitabine-based therapy. Since the clinical trial suggested that physical mass transport properties could describe the variability in delivery of gemcitabine to the tumor cell DNA, we hypothesized that the transport properties could also describe the variable response to and outcome after gemcitabine-based therapies. To test this idea, we correlated the CT parameters with pathological response and survival of patients with PDAC after preoperative chemoradiation [309]. We identified 110 patients who received gemcitabine-based chemoradiation and had evaluable pretherapy CT scans from two prospective clinical trials at our institution for potentially resectable PDAC [310,311]. We observed distinct CT signatures in patients who had complete pathological responses compared with those with poor responses to therapy (Figure 7.6A), and the normalized CT-derived parameter AUC directly correlated with pathological response (Spearman rank-order correlation, –0.30; 95% CI, –0.51 to –0.05; $p = 0.02$; Figure 7.6B). As previously observed by our group [242,309], patients with better grades of

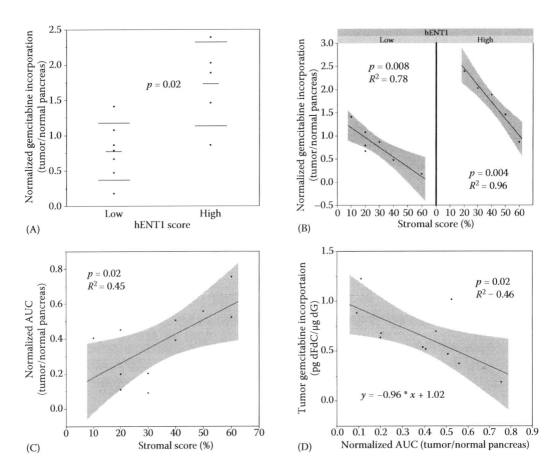

FIGURE 7.5 Correlations between transport properties and gemcitabine incorporation. (A) Normalized gemcitabine incorporation for each patient on the clinical trial of intraoperative gemcitabine infusion during PDAC resection, measured using specimens obtained directly from the tumor. Surgical specimens were scored for hENT1 staining. A significant difference was observed in the normalized gemcitabine incorporation when divided by the hENT1 score (two-tailed t-test; mean and standard deviation indicated by short and long lines, respectively). (B) The stroma amount was scored independently by a pathologist, and gemcitabine incorporation in the tumor and normal pancreas was measured. After accounting for the hENT1 score, a significant inverse correlation was seen with normalized gemcitabine incorporation (linear regression). (C) Pretherapy CT scans of each patient in the clinical trial of intraoperative gemcitabine infusion during PDAC resection were derived, and the normalized CT-derived parameter AUC was plotted against the stromal scores from surgical pathology for the corresponding patient. A direct linear correlation was observed. (D) Normalized AUC was plotted against the measured gemcitabine (dFdC) incorporation into pancreatic tumor cell DNA (expressed relative to deoxyguanosine [dG]). A significant inverse correlation was observed (linear regression), in agreement with the inverse correlation found for the stromal score (B), which directly correlated with the normalized CT parameter AUC (C). The equation indicates how the CT parameter may be used to predict gemcitabine incorporation in future clinical trials. (Reproduced with permission from Koay, E. J. et al., *J. Clin. Invest.*, 124(4), 1525–1536, 2014.)

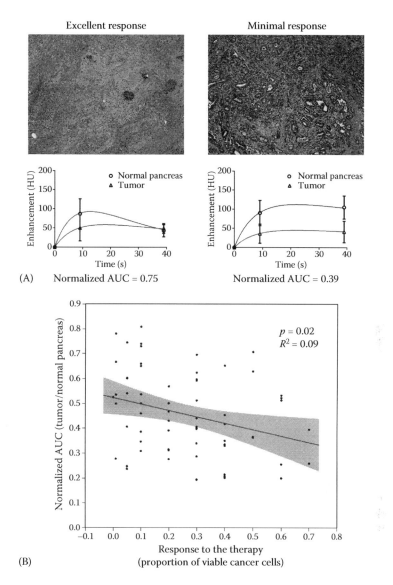

FIGURE 7.6 Correlations between transport properties and response. (A) Representative histology, CT profiles, and normalized AUC values for patients with excellent (trace viable tumor cells) and minimal (approximately 70% viable tumor cells) responses to therapy. The patient with an excellent response had higher normalized AUC than the patient with a minimal response. (B) Normalized AUC was measured from the pretherapy CT scans of the patients who underwent surgery for potentially resectable PDAC in two Phase II clinical trials of preoperative gemcitabine-based regimens [310,311]. The pathological response to therapy was scored by a pathologist. Higher normalized AUC appeared to correlate with better pathological response (linear regression). (Reproduced with permission from Koay, E. J. et al., *J. Clin. Invest.*, 124(4), 1525–1536, 2014.)

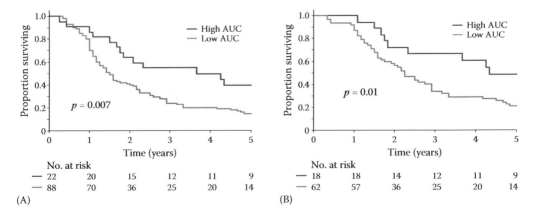

FIGURE 7.7 Correlations between transport properties and survival. (A) Using a partitioning analysis for all 110 patients who received gemcitabine-based therapy in two published Phase II trials for potentially resectable PDAC [310,311], a cutoff of 0.6 was identified for normalized AUC (values greater than 0.6 were considered high; all others were low). This designation separated patients with a good prognosis from those with a poor prognosis on univariate and multivariate analyses. (B) Of the initial 110 patients who received gemcitabine-based therapies for potentially resectable PDAC, 80 underwent curative-intent surgery. When the same cutoff of 0.6 for normalized AUC was applied to these 80 patients, patients with good prognosis were again separated from those with poor prognosis. This finding was significant on univariate and multivariate analyses. (Reproduced with permission from Koay, E. J. et al., *J. Clin. Invest.*, 124(4), 1525–1536, 2014.)

pathological responses (i.e., fewer viable cells after therapy) had improved prognosis, and as response correlated directly with normalized AUC, patients with higher normalized AUC values also had improved prognosis. Multivariate analysis of the 80 patients who underwent resection confirmed that normalized AUC was an independent predictor of overall survival.

An exploratory partitioning analysis identified a cutoff of 0.6 for normalized AUC. Applying this cutoff to the entire cohort of 110 patients showed that patients with high normalized AUC had significantly better outcome than those with low normalized AUC (40% vs. 15% survival rate at five years), independent of whether the patients had curative intent surgery (Figure 7.7A). Additionally, the same cutoff for normalized AUC remained a significant predictor of survival in the 80 patients who underwent surgery, independent of margin status and lymph node involvement (Figure 7.7B).

7.2.4 Assessing Intratumoral Heterogeneity

To reproducibly describe the drug delivery and mass transport properties of PDAC, we developed a volumetric segmentation approach to measure the mass transport properties from CT scans. The images were first registered into a computer program (Pinnacle 9.6); then, the images from three different phases were enhanced, aligned, and analyzed for density. Finally, a volumetric mean was used to calculate the mass transport properties of each structure in the image. This method is easily reproducible and accurate, where there was less than a 5% difference in the average predicted value between three medically trained observers (Figure 7.8). The translational relevance of this work

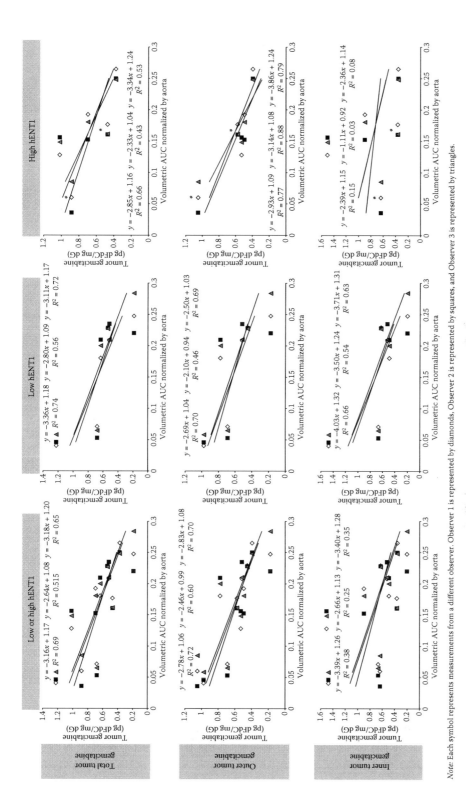

FIGURE 7.8 Degree of heterogeneity of gemcitabine delivery into the inner and outer portions of PDAC tumors, as well as the reproducibility of mass transport measurements by three different observers. Each patient has one measurement for gemcitabine (y-axis) in a particular region of the tumor. The three measurements of volumetric AUC (x-axis) are shown for the different observers. The results are further divided according to the expression of the molecular transporter of gemcitabine, hENT1. dFdC, gemcitabine; dG, deoxyguanosine.

Note: Each symbol represents measurements from a different observer: Observer 1 is represented by diamonds, Observer 2 is represented by squares, and Observer 3 is represented by triangles.

is high, since only routine CT scans, already part of the PDAC diagnosis and treatment protocol, are used.

7.3 IMPLICATIONS

A central goal of clinical oncology is a personalized therapeutic approach for each patient. PDAC is one of the leading causes of death among cancers and has one of the lowest rates of survival. One of the reasons for these poor outcomes is variability and unsuccessful drug delivery. For chemotherapy to kill cancer, the drug must reach the cell by navigating through various barriers presented on multiple levels within the body and tumor microenvironment. As we have confirmed, there are vast physical differences between tumors, and even within an individual tumor, that present barriers to drug delivery. As a result of this variability in drug delivery, there is no one-size-fits-all treatment; consequently, individualized methods must be pursued to achieve better outcomes.

Our method of quantifying mass transport properties can provide insight into a tumor's physical environment, as well as its response to therapy. The CT-derived mass transport properties correlated with the delivery of gemcitabine and with response to therapy. We even found that patient survival could be accurately predicted using their derived properties. The clinical implications are astounding in that this method can be used to give patients a prediction to their response to therapy and chances of survival.

Furthermore, this method could be used in treatment planning. Quantifying the mass transport properties of a tumor requires no extra steps for the patient because only routine pretreatment CT scans are used. From these scans, the transport properties can be derived, which can be done at any institution without complex software or algorithms. These properties can be used to compose a biophysical characterization of the tumor, which can aid physicians in choosing the best treatment plan with a clearer prediction of the probable outcomes.

This work can aid in efforts to improve drug delivery through experiments mimicking the intraoperative drug infusion platform. Researchers could test methods to alter the physical environment of the tumor being removed and observe its effects on drug delivery. This could add valuable information on the effects of the tumor's physical environment to drug delivery and how these barriers can be overcome.

This research provides evidence that drug delivery is affected by the biological properties of pancreatic cancer, leading to heterogeneous drug delivery. Previous work described in the previous chapters has shown that more drug delivery translates to more tumor cells being killed. Combined with the findings of these studies, the conclusion can be made that differences in the response to treatment are due to variations in the physical properties of the tumor.

With further validation and optimization, our CT analysis method may find wide clinical application for both diagnostic and therapeutic planning purposes, as the principles of mass transport can be applied to any human pathological process, as well as a variety of therapeutic agents. The clinical trial design of intraoperative drug infusion is a critical component of our study of mass transport in PDAC, represents a novel platform by which to study mechanisms of targeted drug delivery, and complements our CT analysis.

Combined, the methodologies we developed here and the results we obtained may lead to rational interventions for pancreatic cancer and other solid tumors that improve drug delivery, and thereby extend survival for patients.

Toward the goal of individualization, our unique clinical trial platform can be used to study biological, pathological, and physical correlates of drug delivery in humans. Others have evaluated drug delivery in humans during therapy [312] or measured transport-related changes after chemotherapy [313], but no trial, to our knowledge, has analyzed the mass transport characteristics of the tumor that may have influenced drug delivery. In the development of this clinical trial, we performed extensive calibration, validation, and correlative studies. We also demonstrated that this trial design and methodology were safe. By understanding the factors that influence drug delivery in humans with PDAC, we hope to develop rational interventions that improve therapeutic outcomes. In particular, one can envision using the intraoperative drug infusion clinical trial platform to test methods to alter the physical environment of the tumor, thereby increasing drug delivery, which provides a rationale for future clinical trials that aim to improve outcomes with these strategies.

7.4 CONCLUSIONS

We showed that mass transport properties of an individual human pancreatic cancer and adjacent normal pancreas could be revealed by modeling the changes in enhancement of the tissues at the specific timed phases of the test (i.e., precontrast, arterial, and portal venous). We developed a mathematical model of mass transport of the contrast material based on multiple, systematic measurements of the Hounsfield units of the tissue of interest (i.e., PDAC or normal pancreas) at each time point of the pancreatic protocol CT. With this model, we derived quantitative mass transport parameters that describe influx and efflux rates of contrast, and maximum enhancement of the pancreatic tissues. We also calculated a volumetric area under the model-predicted enhancement curve (VAUC) from the CT scan as a measure of enhancement. This revealed a twofold difference in mass transport properties between normal and malignant pancreatic tissues, which is consistent with the observation that pancreatic tumors are typically hypodense relative to the normal pancreas.

Considering their robust correlations with gemcitabine incorporation, pathological response, and oncologic outcome, CT-derived mass transport parameters represent biophysical markers that may have potentially significant implications for cancer medicine. Further development of diagnostic tests that simultaneously allow radiologic cancer staging and biophysical tumor profiling is warranted. The concept of individually tailored cancer therapy based on biophysical characterization is also supported by our present findings, as patients with good response to therapy appeared to have different physical properties compared with those with poorer responses. Our clinical trial platform of intraoperative drug infusion during resection suggests that the sequential contributions of vascular, extracellular, and cellular transport influence gemcitabine incorporation. Future investigations using this trial platform will aim to better understand these transport mechanisms, validate our findings, and develop rational therapeutic interventions for patients.

CHAPTER 8

Tumor Morphological Behavior and Treatment Outcome

With John Lowengrub

$\mathbf{A}$S DISCUSSED IN CHAPTER 7, pancreatic ductal adenocarcinoma (PDAC) has highly variable clinical outcomes, even though it is generally associated with early distant metastasis (DM) and resistance to chemotherapy and radiation. Accounting for the variability between patients and within tumors could help us to treat this disease better. Some of the key problems with using a purely biological approach are that PDAC tumors are difficult to biopsy due to their location, the cost of molecular characterization such as DNA sequencing, and a dearth of biomarkers that have strong prognostic significance or predictive capability. In collaboration with our mathematical modeling collaborator, Dr. John Lowengrub of the University of California, Irvine, we have proposed using the physical and biological properties of PDAC as an alternative [314].

8.1 INTRODUCTION

Differences in the biophysical properties between individual tumors likely contribute to the variable outcomes of patients [7,8,10,244,307,315]. Toward characterizing these properties, we considered one of the hallmark features of PDAC: extensive desmoplasia (also called stroma). In this regard, preclinical studies suggest that stromal elements of PDAC may impede metastatic spread of the cancer cells [256,257]. A pathology-based study of patients with early-stage PDAC who underwent up-front surgery suggested that patients with more stroma surrounding the tumor had better survival outcomes [316]. Notably, the stroma changes in response to therapy and may be associated with more aggressive cancer biology [317]. Thus, the native and dynamic roles of the stroma remain areas of intense scientific investigation. It is clear that the stroma plays a complex role in PDAC, but there is consensus that clarifying how the stroma influences the cancer cells of PDAC before and during therapy will have important clinical implications that may directly impact patient survival outcomes.

Our work published in the *Journal of Clinical Investigation* illustrated how the stroma may represent a physical barrier to drug delivery in PDAC [8]. This was notable for several reasons. First, the degree of stroma was not sufficient to describe the variation in delivery of the chemotherapy drug gemcitabine by itself. We had to account for how much molecular transporter of gemcitabine was present at the cell surface. This transporter is human equilibrative nucleoside transporter (hENT1), and is variably expressed on the surface of PDAC cells. By showing that these multiscale properties (stroma and molecular transporter) both contributed to gemcitabine delivery, we supported a key principle of mathematical oncology: the problem crosses multiple length scales. Another notable finding of this work was that mass transport properties derived from computed tomography (CT) scans correlated with the delivery of, response to, and outcome after gemcitabine-based therapies [8,307]. These data suggested that properties derived from CT scans could serve as biophysical markers of PDAC, but were unable to provide insight *a priori* into the history of or underlying physical and biological mechanisms related to patient outcomes.

Given this limitation, we think that other physical features of PDAC can be quantified through diagnostic CT scans. Our previous work in breast cancer and glioblastoma suggested that the morphology of the tumors was a reflection of cancer aggressiveness [122,179,318–321]. We specifically studied factors that influenced cancer cell migration and proliferation. These previous investigations were performed at the microscopic scale (microns). To bring the model to the macroscopic scale (centimeters) so that CT scans can be used to characterize the disease, we used our knowledge of the stroma of PDAC. We specifically hypothesized that the stromal elements of PDAC would be a strong influence on PDAC proliferation and migration rates.

8.2 MATHEMATICAL MODEL

We have previously demonstrated [98,179,318,319], using a biophysical theory of tumor morphologic stability [320], that the features of the tumor at the microscopic scale were described by the competition of mechanisms that oppose infiltrative growth (termed "relaxation mechanisms," e.g., cell proliferation) and those that promote infiltrative growth (e.g., cell migration). Motivated by recent studies that suggested that the stroma acts to restrict metastatic spread of PDAC [256,257], we hypothesized that our mathematical theory of tumor growth could be scaled up from the microscopic level to the macroscopic tissue level by considering the stroma as a global relaxation mechanism that strongly influenced gross tumor morphology.

First, we assumed that the major mechanisms responsible for tumor growth, and thus the value of the stability parameter in each patient, are regulated and affected by the stroma through a variety of molecular and physical mechanisms.

1. Macroscopic tumor growth model (for the full model, see [98,179,318,319]): The mechanistic and biological processes of tumor growth and their morphology can be simplified as shown in Equation 8.1:

$$\frac{\partial \varphi}{\partial t} = \nabla \cdot \varphi(\nabla p - \Lambda_M \nabla \sigma) + \Lambda_P \sigma \varphi \qquad (8.1)$$

where the left-hand side represents the rate of change of local concentration φ of live tumor cells, and the right-hand side contains the three mechanisms hypothesized to be responsible for this change. The proliferative "pressure" p (which displaces cells to make room for new cells) and the proliferation rate Λ_P contribute to "relaxation" of tumor morphology into a "ball-like" (high-delta) growing mass. The migration of cells by chemotaxis at a rate Λ_M and with speed proportional to spatial gradient ∇ in local cell nutrient concentration σ are responsible for destabilization of tumor interface morphology. Importantly, the parameter σ has values on the inside and outside of the tumor, and quantifying this difference would be akin to measuring the difference in perfusion between the tumor and surrounding parenchyma.

2. Stability parameter (Λ): The stability parameter is defined as the proliferation rate, Λ_P, divided by the migration rate of cancer cells, Λ_M, that is, $\Lambda = \Lambda_P/\Lambda_M$. The stability parameter can be estimated by surrogates for the proliferative and migration rates of the cancer cells, such as cellularity and cellular morphology (Λ = cellularity/cell axis ratio).

3. Perfusion model: The perfusion within the tumor mass can only occur up to a distance L [$L = (D/\lambda)^{1/2}$] from the tumor border, where D is diffusivity and λ is the cellular uptake rate of molecules that perfuse into the tumor.

Since one of the key model parameters (Λ) describes the stability of the tumor and represents the ratio of factors that influence proliferation and migration, one would expect that a low parameter (i.e., where migration is favored over proliferation) would result in a mixture of cancer cells and stromal cells. On the other hand, a tumor where proliferation dominates would result in relatively low mixing of cancer and stromal cells.

We performed parametric analyses of Λ (low values ranged from 0.2 to 0.25, and high values ranged from 1 to 1.5) to ascertain the effect it would have on the gross morphology of the tumor during our simulations (Figure 8.1). As expected, when the proliferation rate is slower than the migration rate (e.g., a low Λ), the model predicted that the cancer cell clusters will intermingle with the stroma, resulting in an indistinct interface for the tumor (Figure 8.1 shows $\Lambda = 0.2$, left), generating what have been described as "low-mode" instabilities [320]. Conversely, when the proliferation rate is larger than the migration rate (e.g., a high Λ), the model would predict the tumors will grow with a distinct interface as proliferation overcomes any attempts of migrating cells to separate from and leave the main tumor bulk behind as they move into the surrounding normal tissue (Figure 8.1 shows $\Lambda = 1.5$, right). These simulations show direct analogy to the macroscopic features of human PDAC on CT scans.

Further, the computer simulations predicted that tumors with a low Λ (Figure 8.1, left) would have low-mode waves of perfusion throughout the tumor and interface. This may be physically interpreted as a higher degree of stroma that leads to intermingling of stromal

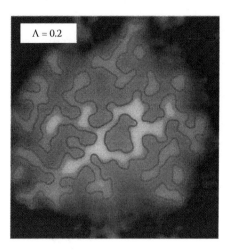
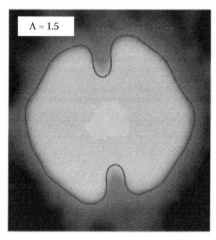

FIGURE 8.1 Simulations of tumor growth. Tumors in which migration rates dominate over proliferation rates will have an indistinct interface between normal pancreas and tumor (left), while tumors in which proliferation rates dominate migration rates will have a distinct interface (right). These morphological patterns can be seen in patient pathology samples and CT scans. (Adapted from Koay, E. J. et al. A visually apparent and quantifiable CT imaging feature identifies biophysical subtypes of pancreatic ductal adenocarcinoma, 2016, submitted.)

and cancer elements throughout. Conversely, tumors with a high Λ parameter would have a higher relative perfusion at the interface of the tumor (Figure 8.1, right). It is suspected that the lower stromal percentage in tumor leads to lower perfusion penetration distances due to a combination of lower diffusivity and higher uptake rates (of nutrients or drugs) by the higher cell content within the tumor. Correspondingly, the higher stability parameter also leads to a roundish distinct tumor interface, as the bulk grows dominated by cell proliferation.

8.3 CLINICAL APPLICATION

These simulations suggest that tumor morphology could provide significant insight into the behavior of pancreatic cancer. Our approach has been to develop reproducible methods to characterize the tumor–pancreas interface on CT scans. One of the methods that we developed is called the delta measurement. This measurement distinguishes the change in enhancement from the pancreatic tumor to the parenchyma of the pancreas (Figure 8.2). Again, this could be interpreted to be the difference in the σ between the parenchyma and tumor. We have also worked with radiologists to visually score the conspicuity of these lesions, where a score of 4 or 5 would correspond to a highly conspicuous (high-delta) tumor, while a score of 3 or less would be an inconspicuous (low-delta) tumor. Based on our model simulations, our hypothesis was that the patients with tumors exhibiting a high delta will have more aggressive biological features and worse clinical outcomes than those with a low delta.

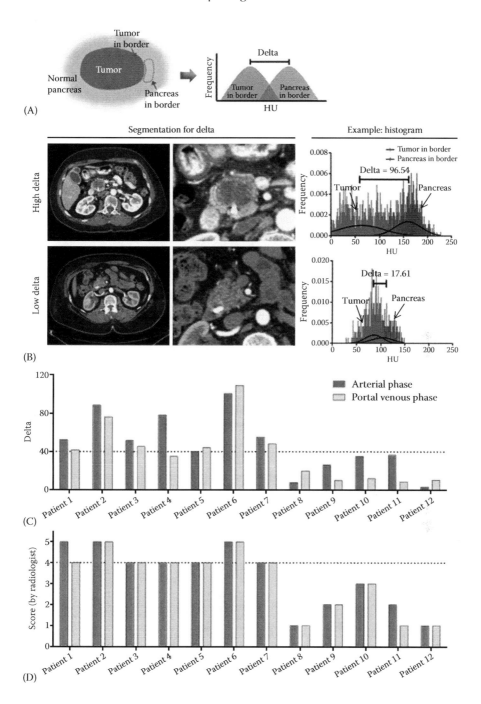

FIGURE 8.2 Measurement of the change in enhancement from normal pancreas to tumor, called the delta measurement. (A) Schematic and (B) patient CT scans. We found that a delta of 40 Hounsfield units (HU) reliably distinguished patients with a conspicuous tumor from those with an inconspicuous tumor (C). Radiologists also scored the conspicuity, and a score of 4 or 5 reliably identified patients with a conspicuous (high-delta) tumor, while those with a score of 3 or less had an inconspicuous (low-delta) tumor (D). (Adapted from Koay, E. J. et al. A visually apparent and quantifiable CT imaging feature identifies biophysical subtypes of pancreatic ductal adenocarcinoma, 2016, submitted.)

When we analyzed the amount of stroma in the tumors of patients who underwent resection for pancreatic cancer, the high-delta tumors had significantly less stroma than those to low-delta tumors (Figure 8.3). We also noted differences in the cancer cells of the tumors (Figure 8.4), whereby cancer cells of high-delta tumors were more elongated than cancer cells of low-delta tumors. The elongated shape indicates a more aggressive biology, as this is associated with mesenchymal biology, which is more likely to metastasize than epithelial biology. Indeed, when we analyzed the clinical outcomes of patients and measured the delta on their CT scans, we observed that patients with a high-delta tumor had worse clinical outcomes than those with a low-delta tumor (Figure 8.5).

The reasons for our clinical observations appear to be rooted in both the intrinsic biology of the cancer cells and the host stromal response to the cancer. The role of the stroma in PDAC disease progression and metastasis has been recognized as an important topic for years, but the evidence has been mixed, likely owing to differences in patient populations, disease stage, treatment approaches, and methods to analyze the stroma. Our mathematical theory of tumor growth and morphology inspired this approach of image analysis of the border interface, and is well supported by studies that demonstrate that the stromal elements of the primary tumor act to restrict metastatic spread. Notably, the model had multiple analogies that were consistent with our quantitative CT imaging feature.

Specifically, the delta measurement reproducibly associated with the amount of stroma in the primary PDAC tumor (Figure 8.3). Moreover, multiple lines of evidence demonstrated that *classifying PDAC according to these imaging features reveals distinctions in the*

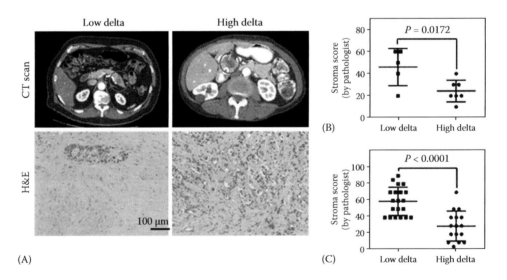

FIGURE 8.3 The tumors with a low delta have different histological appearance than those with a high delta (A). A pathologist quantified the amount of stroma in two cohorts of patients who underwent surgery for pancreatic cancer: (B) 12 patients and (C) 33 patients. Patients with a low delta on the CT scan had significantly higher amounts of stroma than those with a low delta in both cohorts. H&E, hemotoxylin and eosin. (Adapted from Koay, E. J. et al. A visually apparent and quantifiable CT imaging feature identifies biophysical subtypes of pancreatic ductal adenocarcinoma, 2016, submitted.)

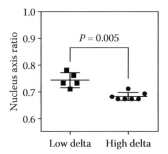 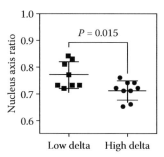

FIGURE 8.4 We measured a nucleus axis ratio of the cancer cells in patients who had surgery for pancreatic cancer from patient histology. The nucleus axis ratio represents the smallest axis divided by the largest axis. The patients who had a low-delta tumor had a significantly higher nucleus axis ratio than those with a high-delta tumor. This indicates these cancer cells were more elongated, and an elongated shape is associated with mesenchymal biology, which is an aggressive phenotype. (Adapted from Koay, E. J. et al. A visually apparent and quantifiable CT imaging feature identifies biophysical subtypes of pancreatic ductal adenocarcinoma, 2016, submitted.)

physical (Figure 8.3) *and biological* (Figure 8.4) *properties of these tumors, as well as in the clinical outcomes for the patients* (Figure 8.5).

This work builds directly on our previous studies that described how to derive mass transport properties from standard-of-care CT scans of patients with PDAC [8,307]. Our previous finding, that a higher volumetric area under the enhancement curve (VAUC) from the CT scan associates with the stroma content, complements the morphological CT feature that we have identified here, but we observed significant overlap between the low- and high-delta groups in terms of VAUC. This likely is due to the heterogeneity in the biology of the disease and stromal response. On univariate analyses, VAUC was associated with survival in our datasets of patients with resectable PDAC. As an exploratory exercise,

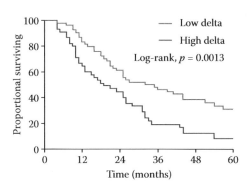

FIGURE 8.5 The survival outcomes of patients differ based on a morphological feature that we can measure on CT scans. We focused on this CT feature because the mathematical model of tumor growth suggested it would help to distinguish tumors in terms of their biological behavior and physical characteristics. (Adapted from Koay, E. J. et al. A visually apparent and quantifiable CT imaging feature identifies biophysical subtypes of pancreatic ductal adenocarcinoma, 2016, submitted.)

we combined the delta classification and the cutoff of VAUC of 0.6 that we used previously to stratify patients [8], and this combined classification appeared to further subdivide the groups of patients (Figure 8.6). Larger datasets will be needed to understand how the morphological characteristics, such as the delta measurement, and physical transport characteristics, such as the VAUC, are connected in PDAC. With further investigation of the differences in the stromal and vascular properties of these imaging-defined groups, one may expect differential effects of antiangiogenic therapy to be correlated with the distinct groups.

These data support our theory of gross morphologic growth of PDAC, which we have previously applied to glioblastoma and breast cancer at the microscopic scale [318,319]. Ongoing work using detailed three-dimensional histological analyses will more precisely estimate model parameters and better determine how the microscopic histological

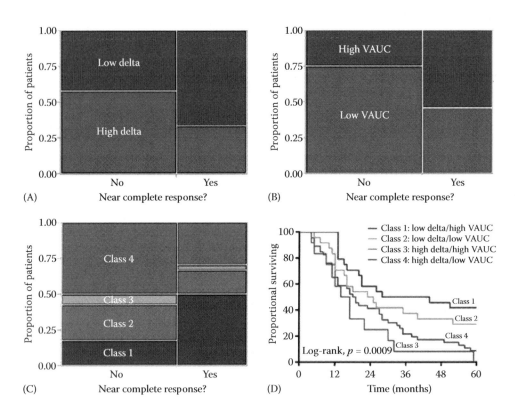

FIGURE 8.6 We have identified two separate imaging features that associate with outcome for PDAC: the VAUC on CT scans and the delta measurement. We explored how the delta class and VAUC associated with the degree of pathological response to preoperative chemoradiation in 106 patients with resectable PDAC (A and B). Interestingly, when we combined the two imaging-based classifiers, we saw that tumors with a low delta and high VAUC are the ones that are most likely to achieve a near-complete pathological response (C). Analyzing the overall survival of these patients using these classifiers shows a difference in prognosis (D). (Adapted from Koay, E. J. et al. A visually apparent and quantifiable CT imaging feature identifies biophysical subtypes of pancreatic ductal adenocarcinoma, 2016, submitted.)

characteristics are related to the macroscopic observations on CT scans and other imaging modalities, like MRI.

The approach of classifying cancer according to physical features that can be measured is advantageous in several ways. First, this approach can use standard-of-care diagnostic CT imaging and can be applied prior to treatment. Most groups use molecular classification approaches to distinguish subtypes of a cancer, including gene expression and mutation profiling [322,323]. The challenge with doing this for PDAC is that this disease can be difficult to obtain enough tissue to perform these expensive studies. Biopsy would likely be limited as a method to assess stromal content due to the intratumoral heterogeneity of PDAC [307]. The assessment of stroma through pathological scoring from surgically resected specimens would not give the information that clinicians need at the time of initial diagnosis, and could not be applied to the 80%–85% of patients with PDAC who present with locally advanced or metastatic disease at diagnosis.

This work builds directly on our previous studies that described how to derive mass transport properties from the standard-of-care CT scans of patients with PDAC [8,307]. Our previous finding that a higher VAUC from the CT scan associates with the stroma content complements the morphological CT feature that we have identified here. We need to investigate how the enhancement measured by VAUC interacts with the delta measurement, as it seems the two are correlated but independent. This will require larger patient datasets. As we carry out these investigations, we are keeping therapy in mind, as described in Chapter 8 about physical biomarkers. With further investigation of the differences in the stromal and vascular properties of these imaging-defined groups, one may expect differential effects of antiangiogenic therapy in these groups.

We are also working with groups around the country to validate these findings. As this occurs, we hope to implement the idea into ongoing clinical trials and learn whether our approach can help to select patients with PDAC for emerging therapies like escalated dose radiotherapy and immunotherapy.

CHAPTER 9

Mechanistic Model of Tumor Response to Immunotherapy

With Geoffrey V. Martin and Eman Simbawa

THE ROLE OF THE immune system in fighting cancer and how immune system inhibition can promote tumor growth has been an area of active research over the past few decades. This relationship between the immune system and the tumor microenvironment is complex, however, as different immune cells can have tumor-suppressive and tumor-promoting properties [324,325]. Furthermore, the interaction between immune cells and cancer cells is likely dependent on a multitude of factors, including tumor antigenicity, individual genetic heterogeneity, prior antigen exposure, tumor vasculature, and nonimmune tumor stromal content [324]. Despite these complexities, the immune system has been shown to have prognostic ramifications in multiple cancer types, with studies demonstrating an association between increased immune cell tumor infiltration on pathologic specimens of colon, endometrial, esophageal, and ovarian cancers and improved disease outcomes [326–329]. Understanding the exact mechanisms of immune system suppression of cancer has led to the development of immune-modulating cancer treatments (immunotherapy), including interleukin-2 (IL-2), vaccine-based therapies, and immune checkpoint inhibitors, with promising results in recent clinical trials [330–335]. In this chapter, we explore general strategies for mathematical modeling of immune system and tumor interaction, as well as specific examples of immunotherapy modeling, which focuses on the application of these computational models in clinical scenarios.

9.1 GENERAL TUMOR–IMMUNE SYSTEM MODEL

Mathematical modeling of tumor–immune system interaction generally begins with simplified assumptions about the dynamics of immune cell killing of tumor cells. In particular, we represent the rate of tumor and immune cell change as coupled ordinary differential equations (ODEs), taking into account generally four key functions: tumor growth, tumor cell kill by immune cells, immune cell recruitment to or proliferation within the tumor, and immune cell inactivation, death, or anergy. These models are therefore represented by the following equations:

$$\frac{d\rho}{dt} = g(\rho,\psi) - h(\rho,\psi) \tag{9.1}$$

$$\frac{d\psi}{dt} = x(\rho,\psi) - y(\rho,\psi) \tag{9.2}$$

where ρ represents the number of tumor cells, ψ represents the number of immune cells in the tumor, and t is time. The terms f, g, x, and y represent functions that depend on ρ and ψ and which stand for tumor cell proliferation (g), tumor cell death induced by immune cells (h), immune cell recruitment or proliferation (x), and immune cell inhibition, death, or anergy (y). These equations are very general in that they can account for multiple types of tumor or immune cells, and their functions of proliferation or death are only functions of time-dependent values in ρ and ψ, respectively. The functions of g, h, x, and y are introduced here as very generalized representations, but can model their biologic meanings in a variety of ways, such as logistic or Gompertzian tumor growth, Michaelis–Menten immune cell recruitment, and saturation kinetic models of immune cell killing of cancer cells. These simple models take into account only the kinetics of tumor and immune cell interactions, but, depending on the choice of the functions on the right-hand side of Equations 9.1 and 9.2, this could still lead to a significant number of parameters that require specification. For example, a Gompertz function of tumor growth has three parameters, and logistic growth or Michaelis–Menten models generally have two parameters. An example tumor–immune cell system with logistic tumor growth, linear tumor cell killing by immune cells, Michaelis–Menten immune cell recruitment from the circulation, and constant immune cell death, inactivation, or anergy by tumor cells would be represented as

$$\frac{d\rho}{dt} = r\rho\left(1 - \frac{\rho}{K}\right) - a\cdot\psi \tag{9.3}$$

$$\frac{d\psi}{dt} = \frac{V(\psi_b - \psi)}{K_m + (\psi_b - \psi)} - b\cdot\rho \tag{9.4}$$

where r is the maximum rate of tumor growth, K is the carrying capacity of the tumor, a is the rate of tumor death by immune cells, V is the maximum rate of immune cell recruitment from circulation, ψ_b is the number of immune cells in the circulation, K_m is the half maximum value of immune cell recruitment from the circulation, and b is the rate of immune cell death or inactivation by tumor cells.

Specification of the exact functions on the right-hand side of Equations 9.1 through 9.4 and their parameters is often performed by fitting their functions to combinations of *in vitro* data for tumor cell kill by immune cells, *in vivo* data for intratumoral recruitment and proliferation of immune cells, and clinical patient data for estimation of tumor growth kinetics. Incorporation of additional immune system–specific variables can be done in a similar way through the inclusion of other ODEs, taking into account the production and destruction of cytokines, antigen presentation, interactions between various immune cells (e.g., regulatory T cells), tumor vasculature, and so forth. Multiple compartments of the immune system (tumor, blood, lymph node, spleen, etc.) can also be modeled with similar techniques. A more sophisticated model taking into account the presence of regulatory immune cells (R) and cytokines (C) would produce a system of equations such as

$$\frac{d\rho}{dt} = r\rho\left(1 - \frac{\rho}{K}\right) - a \cdot \psi \tag{9.5}$$

$$\frac{d\psi}{dt} = \frac{V(\psi_b - \psi)}{K_m + (\psi_b - \psi)} + d \cdot C - b \cdot \rho \tag{9.6}$$

$$\frac{dR}{dt} = c \cdot \psi - e \cdot C \tag{9.7}$$

$$\frac{dC}{dt} = d \cdot \psi \tag{9.8}$$

where ρ, ψ, r, K, a, V, ψ_b, K_m, and b represent the same parameters as in Equations 9.3 and 9.4, and c, d, e, and f represent additional parameters modeling the interaction effects of ρ, ψ, R, and C. These equations are graphically depicted in Figure 9.1. Overall, the complexity

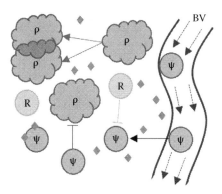

FIGURE 9.1 The figure depicts a sample tumor–immune cell environment represented by mathematical system of equations (Equations 9.5 through 9.8). Effector immune cells (ψ) are represented to be present within the tumor environment and infiltrating from a blood vessel (BV). Other immune cells, including regulatory immune cells (R), are represented as having inhibitory action on effector immune cells. Tumor cells (ρ) are killed by effector immune cells and also shown to be dividing. Diamonds represent sample cytokines (C) within the environment.

of the interactions between tumor and immune cells allows for the use of myriad modeling techniques, and understanding which models or parameters are most appropriate for a given situation requires diligence, but can potentially help identify relevant information and predict treatment response, as we will see in this chapter.

9.2 MODELING IMMUNE MODULATING THERAPIES IN CANCER

After an appropriate model has been chosen for the tumor–immune system interaction, the intervention of immunotherapy can be added to the equations for analysis of sensitive parameters and predictions of patient tumor response. These predictions then need to be validated against laboratory or clinical response data for confirmation of the model's accuracy. Furthermore, a sensitivity analysis of the parameter values used in the model needs to be investigated, as the sensitivity of parameters may help determine other potential therapeutic targets or identify significant sources of variability in the datasets and models.

In general, immunotherapies encompass any therapeutic strategy aimed at modulating the immune system to gain a desired clinical end. In cancer, immunotherapy strategies have included vaccination, generalized immune system activation (e.g., IL-2), and targeted immune checkpoint inhibitors (e.g., anti-programmed cell death protein-1 [PD-1] antibodies), as shown in Figure 9.2. These therapies emphasize the importance of relevant model selection, as they can have different effects on individual components of the immune system. For example, dendritic cell processing of tumor antigens may be important for adequate vaccination response modeling, but may be less relevant for checkpoint inhibitors only targeted at T-cell surface receptors. Finding not only biologic variables that are relevant to immunotherapy response but also ones that are measurable or estimated from other studies remains another challenging task in quantitative modeling of immune-based therapies. Despite these challenges, multiple studies have investigated the role of these interventions and correlated their findings to clinical data. To exemplify the role of mathematical modeling in immunotherapy, we briefly describe some successes of prior

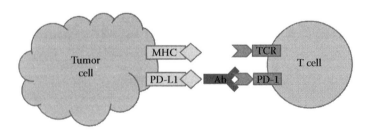

FIGURE 9.2 Depiction of immune checkpoint antibody therapy. A cellular scenario depicts anti-PD-1 antibody therapy. Activation of T cells generally relies on the interaction between the major histocompatibility complex (MHC) molecule on the tumor cell and the T-cell receptor (TCR) on the T cell. Expression of PD-L1 on tumor cells with binding to PD-1 on T cells causes immune cell inhibition. The anti-PD-1 antibody blocks the interaction between PD-L1 expressed on tumors and PD-1 expressed on immune cells, thereby decreasing the immune suppression of the tumor cells. Ab, antibody.

investigations in this area, as well as present a model of immunotherapy adapted from our previously detailed chemotherapy models in the rest of this chapter.

9.2.1 Immunotherapy Modeling Examples

Given the success of vaccination over the past century as a therapy to prevent deadly infections, there has been significant interest in translating this success to cancer therapy, especially as the role of the immune system in fighting cancer continues to evolve. One of the most successful uses of a vaccine therapy in cancer has been that of the Bacillus Calmette–Guérin (BCG) vaccination for superficial, non-muscle-invasive bladder cancer. After transurethral resection, the addition of intravesical BCG has been shown to decrease the chances of bladder cancer recurrence at 12 months compared with no further treatment (odds ratio 0.30) [336]. Despite its success in bladder cancer, not all patients derive benefit from BCG treatment, and its use can be associated with side effects, such as increased urinary frequency, dysuria, hematuria, and generalized malaise [336]. To better understand the mechanisms of BCG administration in patients, Bunimovich-Mendrazitsky et al. used a simple mathematical model to show that full tumor eradication or recurrence is dependent on initial conditions [337]. Their model is an extension of the general framework we presented above utilizing coupled ODEs and parameters derived from the literature for tumor growth rate, immune cell death rate, infection rate of tumor cells by BCG, and so forth. Their mathematical model consisted of a system of 4 ODEs and 11 parameters, and their results showed that tumor eradication was dependent on the initial growth rate of tumor and the treatment rate. Given these conditions, the remaining tumor after transurethral resection could be completely eradicated or undergo continued growth with recurrence in the range of clinical treatment regimens. Comparing the predicted results of response rates over a simulated patient population led to very similar complete response rates (80%) compared with published clinical trials of BCG administration [338].

Another example of a mathematical model of vaccination therapy in cancer was performed by Kronik et al. in the setting of prostate cancer [339]. Their study focused on the mathematical dynamics of prostate cancer vaccination and prostate-specific antigen (PSA) response. Their model took into account multiple compartments associated with immune system activation and effector cell action. They modeled dendritic cell maturation in the dermis after subcutaneous injection, activated dendritic cell migration to lymph nodes, stimulation of effector (cytotoxic T cells) and inhibitory (regulatory T cells) immune cells within the lymph node, and finally, immune cell killing of prostate cancer cells. Their model consisted of 7 coupled ODEs and specification of 15 parameters. Validation of this model with clinical data showed accurate prediction of PSA levels in 12 of 15 patients with an R^2 of 0.972 between predicted and actual PSA values after vaccine administration. They further predicted with their model that alteration to the vaccination schedule on an individual basis could enhance the efficacy of vaccination therapy, and that these alterations varied by individual patient for maximum therapeutic effect, although these predictions were not clinically validated.

Similar to modeling vaccine strategies, systems of ODEs can be used to model systemic immune activating agents. One example of this strategy was an investigation by de Pillis et al. that studied the role of interleukin-21 (IL-21) as an immune stimulating agent for anticancer therapy [340]. IL-21 has been found to exhibit antitumor effects by increasing immune cell–mediated killing of tumor cells, inducing lasting antitumor immune cell memory, and reducing angiogenic and metastases in various tumors [341,342]. One important proposed mechanism for IL-21 efficacy is the transition from an innate natural killer (NK) cell response to a more effective and specific cytotoxic T-cell (CD8+) antitumor response. To model these interactions, de Pillis et al. used six ODEs representing the IL-21 concentration in blood, population dynamics of NK cells in the spleen, population dynamics of specific antitumor CD8+ T cells in the lymph nodes, an element facilitating CD8+ T-cell memory, a cytotoxic protein affecting tumor lysis, and tumor mass. Their model consisted of 21 parameters that were estimated by fitting to experimental murine data or obtaining biologically relevant estimates from the literature. The model was able to predict the growth patterns of multiple types of tumors after receiving a variety of IL-21 dosing schedules, and predicts that different amounts of IL-21 lead to tumor eradication based on tumor mass and tumor antigenic properties. These examples of immunotherapy in cancer show that simple systems of ODEs are able to predict experimental or clinical results, but their full utility in tailoring treatment to individual patients remains a task for future validation.

9.2.2 Modeling Immune Checkpoint Inhibitors

In this section, we extend our prior work on modeling chemotherapy to the realm of immunotherapies. We focus on modeling immune checkpoint inhibitor antibodies due to their success in recent clinical trials, although they can be extended to exogenous molecular immunotherapies in general. Briefly, immune checkpoint inhibitors attempt to block the interaction between immune-inhibiting ligands expressed on tumor cells and their binding counterparts on immune cells. Once the tumor ligands are bound to these proteins expressed on the immune cells, they begin a cascade of intracellular events that renders the immune cells ineffective at killing tumor cells. Tumors have a multitude of ways that they can inhibit immune cell killing, but these immune checkpoint pathways represent a prominent mode in some types of cancers. Two of the most clinically relevant immune checkpoint pathways are those associated with cytotoxic T lymphocyte–associated protein-4 (CTLA-4) and PD-1 [343,344]. Blocking the interaction between the tumor ligands specific for CTLA-4 or PD-1 (cluster of differentiation [CD] 80/86 or programmed death ligand-1 [PD-L1], respectively) on immune cells with anti-CTLA4 or anti-PD-1 antibodies has shown clinical responses in colorectal, lung, melanoma, urothelial, and renal cell cancers [293,330,345–347].

The effects of immune checkpoint inhibitors for cancer treatment have been incorporated into mathematical models in a variety of ways. For example, in a study by Wilkie and Hahnfeldt, immune checkpoint blockade was modeled as increasing the parameters of immune cell recruitment to and activity within tumors [348]. This assumption is consistent with *in vitro* and *in vivo* data showing increased tumor infiltration by immune cells, immune cell movement, and immune cell activity against tumors with increasing activation [349,350]. The results from their modeling demonstrate that increasing these parameters improves the chances of immune-induced tumor elimination, although this remains dependent on tumor growth rates and tissue carrying capacity.

The model of immune checkpoint inhibitor immunotherapy we propose in this chapter expands upon this work by replacing the generalized chemotherapy model of tumor cell mass represented by Equations 4.4 and 4.5. The model considers immune cell presence necessary for tumor cell kill and adds an ODE to model effector immune cell concentration within the tumor. Even though many of the immune checkpoint inhibitor antibodies used in clinical trials have been bound to the receptor on immune cells and not directly on the tumor cells, the site of effect happens at the tumor cell interface, allowing us to modify our prior equations of cytotoxic chemotherapy to apply to immune checkpoint inhibitor antibodies (Figure 9.2). These assumptions lead to the following set of coupled ODEs:

$$\frac{d\rho}{dt} = \alpha \cdot \rho - \lambda_p \cdot \rho \cdot \psi \cdot \int_0^t \lambda \cdot \rho \cdot dt' \tag{9.9}$$

$$\frac{d\psi}{dt} = \Lambda_\psi \cdot \frac{d\rho}{dt} \tag{9.10}$$

where ρ and ψ represent the number of cancer cells and effector immune cells, respectively. Moreover, α is the proliferation rate of tumor cells, λ_p is the specific death rate of cancer cells, λ is the blocking of immune inhibitory ligands on cancer cells via immunotherapy antibodies, and $\Lambda\psi$ is the number of immune cells that are removed or anergic after killing cancer cells. This model assumes that tumor growth is exponential, that immune cell killing of tumor cells is dependent on the amount of immunotherapy antibody present in the tumor, and that immune cell presence within the tumor is primarily dependent on the intratumoral immune cells at the time of immunotherapy administration. If we integrate Equation 9.10, this yields

$$\psi = \psi_0 + \Lambda_\psi \cdot (\rho - \rho_0) \tag{9.11}$$

where ψ_0 and ρ_0 are the number of immune cells and tumor cells at $t = 0$. Substituting Equation 9.11 into Equation 9.9 yields the following tumor cell dynamics:

$$\frac{d\rho}{dt} = \alpha \cdot \rho - \lambda_p \cdot \rho \cdot (\psi_0 + \Lambda_\psi \cdot (\rho - \rho_0)) \cdot \int_0^t \lambda \cdot \rho \cdot dt' \tag{9.12}$$

To compare the model with clinical data, we replace the amount of tumor cells ρ by a proportion of the original tumor volume $\rho' = \rho/\rho_0$, express $\Lambda_{\psi\rho} = \Lambda_\psi \cdot \rho_0/\psi_0$, and assume after a short transient time that the tumor has reached a saturation of the immunotherapy drug $\int_0^t \lambda \cdot \rho \cdot dt' = \sigma$, and that the effect of immune cell killing of cancer cells is a constant dependent on $\lambda_p \cdot \psi_0 \cdot \sigma = \mu$. Replacing these terms in Equation 9.12 yields

$$\frac{d\rho'}{dt} = \alpha \cdot \rho' - \rho' \cdot [1 + \Lambda_{\psi\rho} \cdot (\rho' - 1)] \cdot \mu \tag{9.13}$$

This simplified model of tumor response to immunotherapy is dependent on three key parameters: α, $\Lambda_{\psi\rho}$, and μ. To investigate the predictions of this model, we focus on two scenarios of immunotherapy response in the setting where each immune cell can kill one tumor cell before becoming inactive or dying (i.e., $\Lambda_{\psi\rho} = \rho_0/\psi_0$). Suppose that the number of immune cells within the tumor is much larger than the number of tumor cells, implying $\Lambda_{\psi\rho} = 0$. Solving Equation 9.13 in this setting will produce an exponential response of tumor cells, $\rho' = e^{(\alpha - \mu) \cdot t}$, implying that tumor response will only be dependent on the rates of immune cell killing versus tumor cell growth, and that the amount of tumor cells will either decay or grow exponentially. If instead we assume that the number of cancer cells is greater than the number of immune cells in the tumor, $\Lambda_{\psi\rho} > 1$, then Equation 9.13 predicts

$\rho' > \left(\frac{\alpha}{\mu} - 1\right) \cdot (\Lambda_{\psi\rho})^{-1} + 1$ if $\frac{d\rho'}{dt} < 0$ (i.e., tumor response) and $\rho' < \left(\frac{\alpha}{\mu} - 1\right) \cdot (\Lambda_{\psi\rho})^{-1} + 1$ if $\frac{d\rho'}{dt} > 0$ (i.e., tumor progression). This means that the number of tumor cells will always be >0 and that the tumor response will be dependent on a combination of the three parameters. For example, assume $\frac{\alpha}{\mu} = 0.5$ and $\Lambda_{\psi\rho} = 2$; then ρ' will always be greater than 0.75, or the largest response of the tumor will be a 25% reduction in the initial tumor volume. Alternatively, if $\frac{\alpha}{\mu} = 5$ and $\Lambda_{\psi\rho} = 2$, then ρ' will show progression with tumor growth of 300%. These scenarios are depicted in Figure 9.3, and a phase diagram showing tumor response as $\Lambda_{\psi\rho}$ and μ change is depicted in Figure 9.4.

To validate this simplified model of immune checkpoint inhibitor response, we fit Equation 9.13 to anti-PD-1 response data in melanoma from Topelian et al. [287] to estimate $\Lambda_{\psi\rho}$ and μ assuming a tumor doubling time of 100 days ($\alpha = 0.0069$) [351]. These fits to clinical response data in 15 patients from the trial are depicted in Figure 9.5, which *exemplifies the ability of Equation 9.13 to model disease progression on treatment, stable disease, and partial or near-complete treatment response*. Ultimately, our goal is to use this modeling strategy for future prediction of individualized patient response to immune checkpoint inhibitors based on biologically measurable variables, similar to the success of our chemotherapy modeling and immunotherapy models in the literature.

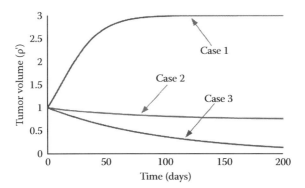

FIGURE 9.3 Simulated tumor response to immunotherapy for different parameter values. Tumor volume (ρ') is normalized to starting size at the initiation of immunotherapy (time = 0). Case 1 depicts simulated tumor progression with $\dfrac{\alpha}{\mu} = 5$ and $\Lambda_{\psi\rho} = 2$. Case 2 depicts partial tumor response with $\dfrac{\alpha}{\mu} = 0.5$ and $\Lambda_{\psi\rho} = 2$. Case 3 depicts complete tumor response with $\dfrac{\alpha}{\mu} = 0.5$ and $\Lambda_{\psi\rho} = 0$.

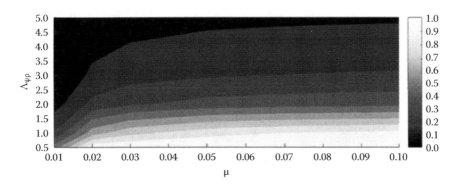

FIGURE 9.4 Phase diagram of simulated tumor response stratified by model parameters, $\Lambda_{\psi\rho}$ and μ. Each point on the figure depicts the simulated tumor response at a given $\Lambda_{\psi\rho}$ and μ with constant tumor doubling time ($\alpha = 0.0069$). Tumor response is encoded by grayscale, with black representing no tumor response and white representing complete tumor response. The color bar on the right shows the full tumor response color spectrum.

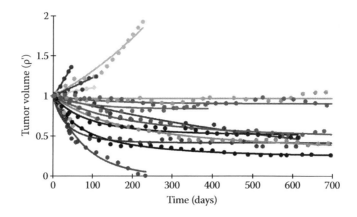

FIGURE 9.5 Plot of immunotherapy model fit to clinical response data in patients with melanoma. Fifteen patient immunotherapy response curves were extracted from [29], with data depicted by circles. An immunotherapy model (Equation 9.13) was fit to these curves assuming a tumor doubling time of 100 days ($\alpha = 0.0069$) to determine $\Lambda_{\psi\rho}$ and μ for each patient. Fits are represented by solid lines; each line represents an individual patient.

9.3 CAVEATS IN IMMUNOTHERAPY MODELING

This chapter has focused on the general mathematical form of immunotherapy modeling using coupled ODEs and featured some of the successes of its application, including ongoing computational modeling simulations, but the complexity of the immune system and its interactions with tumor cells leads to the possibility of near-endless models. Even after choosing the appropriate interactions to model with coupled ODEs, including the relevant aspects of the immune system, there are often dozens of parameters that need to be determined either by fitting to individual patient data, derived from *in vitro* or *in vivo* experiments, or estimated. Even validated mathematical models can show extreme sensitivity to these parameter values, with tumor responses varying widely. For example, de Pillis et al. showed that tumor response could vary between –50% and +50%, depending on a –1% or +1% change in a T-cell efficacy parameter [340]. These extremely sensitive parameters may identify biologic mechanisms with significant therapeutic potential, but also identify areas where even small measurement errors can lead to significant differences in tumor response, limiting their application for quantitative clinical predictions. Another concern with applying complex immunotherapy models in the clinical setting is obtaining patient-specific biologic data. Many immunotherapy models rely on things such as immune cell killing efficacy and antigenicity of tumor antigens, which may be difficult to determine on an individual basis without invasive or expensive testing. *Our approach to immunotherapy modeling has therefore been to start with the simplest possible models and most important biologically relevant parameters for the prediction of clinical response, with a focus on model robustness to address some of these issues.*

9.4 CONCLUSIONS

In conclusion, the field of mathematical modeling of immunotherapies remains a promising one. Despite the complexities of the immune system, simple models using ODEs and limited biologic variables help provide insights into immune modulating therapies. These models provide predictions on treatment response, which may be useful to maximize treatment efficacy in individualized patients in the near future. Additional intricacies, including stochastic variables and equations, agent-based modeling, and spatial dependencies, can be combined with the proposed framework of this chapter to improve model predictability and more fully understand the interactions between tumors and the immune system. Identification of relevant biomarkers through histologic analysis, diagnostic imaging, or blood samples will also help to improve individualized parameter estimation within the models. As our understanding of cancer and the immune system grows, treatments designed to harness the power of the immune system in fighting cancer will also continue to improve, further supporting the need for rigorous, quantitative models of individualized therapy response.

CHAPTER 10

Perspectives on Physical Oncology and Future Directions

CURRENTLY, THE MAINSTAYS OF treatment for most forms of cancer are chemotherapy, radiation, and surgery. Unfortunately, many patients fail these conventional treatments, and it is difficult for physicians to understand how much tumor could be affected (or killed eventually) by the therapy in any given patient. In just the past half decade, we and our colleagues have clearly illustrated the need for a physical oncology–based approach to treating cancer, by viewing cancer as a combination of physical and biological problems, rather than a purely biological one. Fundamental physical processes in tumors, such as the movement of nutrients, delivery of drug molecules, and exchange of mechanical forces between cancer cells and the surrounding tissue, depend on many biological processes, including the growth of blood vessels, the formation of fibrous collagen matrix, and cancer cell proliferation. These biological processes exert physical forces on the cancer cells and ultimately influence the growth patterns of the tumor and its response to treatment. By developing mathematical models to describe these growth patterns and tumor responses to therapies, we have recast the biological problem into a bioengineering one.

As described throughout the book, current treatments are ineffective and physicians are unable to accurately predict the outcomes to particular treatments. It is time for an individualized approach to cancer treatment; each tumor is unique and should be treated as such. Physical oncology enables experts to account for multiple scales of drug delivery, from the cellular scale to the tissue scale to the effect on the whole body. This approach also allows physicians to identify potential barriers to drug delivery on any scale and create treatment plans to overcome them.

Our approach is clinically relevant because we use quantitative measurements from standard-of-care diagnostics to render our mathematical predictions. Hence, physicians at the forefront of treating cancer patients will have quantitative tools to assist them in designing therapeutic regimens for an individual patient, beyond current subjective and heuristic assessments. In fact, the idea of using mathematics and physics to optimize processes is highly successful in many other industries, including finance, oil and gas, computers, and the automotive industry; so we believe similar quantitative and objective approaches can benefit oncology.

We will next focus on translating the models to clinical use. We will develop practical mathematical tools that can be used in the clinic by physicians to predict treatment outcome for each individual patient prior to actual treatment. Since these tools are derived from fundamental principles of mass transport, they can be broadly applicable to the clinical sciences. More importantly, information to build and use these tools can be directly assessed from computed tomography scans, patient tissue analyses, MRI, mammography, and other noninvasive or minimally invasive procedures that the patient would normally receive. With these quantitative tools, oncologists will be able to determine more effective, patient-specific drug treatment strategies (i.e., individualized medicine), such as the amount, frequency, and delivery platform of drug and the need for ancillary non-drug-based treatments.

More research is needed to translate mathematical models to patients in the clinic. In particular, we have scientific and practical problems to overcome. For example, we need to understand these physical phenomena in tumors in the contexts of emerging therapeutic strategies like targeted drugs, immunotherapy, metabolic therapy, and nanoparticles. In collaboration with an interdisciplinary network of colleagues from across the nation and around the globe, we are refining our measurement techniques and modeling approaches to study interrelated biological processes such as metabolism and immune responses in cancer. Our current efforts with functional imaging may help to address this challenging problem. Another major challenge that lies ahead will be to make our measurement techniques and models accessible to the entire scientific and medical community so that they can be implemented for any patient, anywhere. We are developing partnerships with industry to help realize this goal.

We have endeavored a multifaceted approach to tackle the major challenges facing oncology. With immunotherapy, we are developing quantitative methods to noninvasively assess the immunological microenvironment of lung cancer, colon cancer, and pancreatic cancer. We are coupling these efforts with clinical trials of immunotherapy. Here, we anticipate that combining our physical measurements with biological assessments of immunological properties of cancer will be critical to improving the treatment efficacy of various immunotherapies, including vaccines and checkpoint blockade. Additionally, we are taking our initial findings with cytotoxic therapies like chemotherapy and radiotherapy in colorectal cancer and pancreatic cancer to the next steps through ongoing trials at MD Anderson Cancer Center and elsewhere. For example, we are developing new imaging techniques with MRI in pancreatic cancer to measure additional physical properties of these tumors and identify how these properties further differentiate aggressive from less aggressive disease beyond our current computed tomography–based measurements. These insights will allow us to stratify patients into different prognostic groups, and by doing so, we can make informed decisions about how to treat these patients.

This type of research lies at the cusp of multiple fields, including medicine, biology, engineering, mathematics, and physics. We work with experts in each of these areas to help us make the measurements and develop the tools we need to test our models. This highly integrated and collaborative work is unique in the area of oncology and has yielded new insights into tumorigenesis, the emergence of treatment resistance, the development of toxicity, and methods to improve therapeutic efficacy.

It can be expected that this research effort driven by quantitative sciences, including mathematics, physics, and engineering, will help oncologists determine more effective drug treatment strategies that individualize the amount, frequency, and delivery platform of drugs, and assess the need for ancillary non-drug-based treatments. In this manner, we hope to improve the outcomes of patients with cancer.

References

1. Shaffer DR, Scher HI. Prostate cancer: A dynamic illness with shifting targets. *Lancet Oncology* 2003;4(7):407–14.
2. Bissell MJ, Radisky D. Putting tumours in context. *Nature Reviews Cancer* 2001;1(1):46–54.
3. Murphy SL, Xu J, Kochanek KD. Deaths: Final data for 2010. National vital statistics reports: From the Centers for Disease Control and Prevention, National Center for Health Statistics, National Vital Statistics System. *National Vital Statistics Reports* 2013;61(4):1–117.
4. The state of cancer care in America, 2014: A report by the American Society of Clinical Oncology. *Journal of Oncology Practice* 2014;10(2):119–42.
5. Pasquier J, Magal P, Boulange-Lecomte C, Webb G, Le Foll F. Consequences of cell-to-cell P-glycoprotein transfer on acquired multidrug resistance in breast cancer: A cell population dynamics model. *Biology Direct* 2011;6:5.
6. Daniel C, Bell C, Burton C, Harguindey S, Reshkin SJ, Rauch C. The role of proton dynamics in the development and maintenance of multidrug resistance in cancer. *Biochimica et Biophysica Acta* 2013;1832(5):606–17.
7. Pascal J, Bearer EL, Wang Z, Koay EJ, Curley SA, Cristini V. Mechanistic patient-specific predictive correlation of tumor drug response with microenvironment and perfusion measurements. *Proceedings of the National Academy of Sciences of the United States of America* 2013;110(35):14266–71.
8. Koay EJ, Truty MJ, Cristini V, Thomas RM, Chen R, Chatterjee D et al. Transport properties of pancreatic cancer describe gemcitabine delivery and response. *Journal of Clinical Investigation* 2014;124(4):1525–36.
9. Edgerton ME, Chuang YL, Macklin P, Yang W, Bearer EL, Cristini V. A novel, patient-specific mathematical pathology approach for assessment of surgical volume: Application to ductal carcinoma in situ of the breast. *Analytical Cellular Pathology (Amsterdam)* 2011;34(5):247–63.
10. Pascal J, Ashley CE, Wang Z, Brocato TA, Butner JD, Carnes EC et al. Mechanistic modeling identifies drug-uptake history as predictor of tumor drug resistance and nano-carrier-mediated response. *ACS Nano* 2013;7(12):11174–82.
11. Hosoya H, Dobroff AS, Driessen WH, Cristini V, Brinker LM, Staquicini FI et al. Integrated nanotechnology platform for tumor-targeted multimodal imaging and therapeutic cargo release. *Proceedings of the National Academy of Sciences of the United States of America* 2016;113(7):1877–82.
12. Humphrey LL, Helfand M, Chan BK, Woolf SH. Breast cancer screening: A summary of the evidence for the U.S. Preventive Services Task Force. *Annals of Internal Medicine* 2002;137(5 Pt 1):347–60.
13. Merajver SD, Iniesta MD, Sabel MS. Inflammatory breast cancer. In *Diseases of the Breast, 4th edition*, ed. Harris JR, Kippman ME, Morrow M et al. Philadelphia: Lippincott Williams & Wilkins; 2010, pp. 762–73.
14. Warren LE, Guo H, Regan MM, Nakhlis F, Yeh ED, Jacene HA et al. Inflammatory breast cancer: Patterns of failure and the case for aggressive locoregional management. *Annals of Surgical Oncology* 2015;22(8):2483–91.

15. van Uden DJ, van Laarhoven HW, Westenberg AH, de Wilt JH, Blanken-Peeters CF. Inflammatory breast cancer: An overview. *Critical Reviews in Oncology/Hematology* 2015; 93(2):116–26.

16. Mason O, Verwoerd M. Graph theory and networks in biology. *IET Systems Biology* 2007; 1 (2):89–119.

17. Kremling A, Saez-Rodriguez J. Systems biology—An engineering perspective. *Journal of Biotechnology* 2007;129(2):329–51.

18. Hanahan D, Weinberg RA. The hallmarks of cancer. *Cell* 2000;100(1):57–70.

19. Kreeger PK, Lauffenburger DA. Cancer systems biology: A network modeling perspective. *Carcinogenesis* 2010;31(1):2–8.

20. Aldridge BB, Burke JM, Lauffenburger DA, Sorger PK. Physicochemical modelling of cell signalling pathways. *Nature Cell Biology* 2006;8(11):1195–203.

21. Chiang AW, Liu WC, Charusanti P, Hwang MJ. Understanding system dynamics of an adaptive enzyme network from globally profiled kinetic parameters. *BMC Systems Biology* 2014;8:4.

22. Marino S, Hogue IB, Ray CJ, Kirschner DE. A methodology for performing global uncertainty and sensitivity analysis in systems biology. *Journal of Theoretical Biology* 2008;254(1):178–96.

23. Gilbert SF. *Developmental Biology*. 8th ed. Sundarland, MA: Sinauer Associates; 2006.

24. Alon U. Network motifs: Theory and experimental approaches. *Nature Reviews Genetics* 2007;8(6):450–61.

25. Lauffenburger DA. Cell signaling pathways as control modules: Complexity for simplicity? *Proceedings of the National Academy of Sciences of the United States of America* 2000;97(10):5031–3.

26. Nochomovitz YD, Li H. Highly designable phenotypes and mutational buffers emerge from a systematic mapping between network topology and dynamic output. *Proceedings of the National Academy of Sciences of the United States of America* 2006;103(11):4180–5.

27. Hofmeyr JS, Cornish-Bowden A. Regulating the cellular economy of supply and demand. *FEBS Letters* 2000;476(1–2):47–51.

28. Sauro H. The computational versatility of proteomic signaling networks. *Current Proteomics* 2004;1(1):67–81.

29. Sauro HM, Kholodenko BN. Quantitative analysis of signaling networks. *Progress in Biophysics and Molecular Biology* 2004;86(1):5–43.

30. Behar M, Dohlman HG, Elston TC. Kinetic insulation as an effective mechanism for achieving pathway specificity in intracellular signaling networks. *Proceedings of the National Academy of Sciences of the United States of America* 2007;104(41):16146–51.

31. Del Vecchio D, Ninfa AJ, Sontag ED. Modular cell biology: Retroactivity and insulation. *Molecular Systems Biology* 2008;4:161.

32. Goncalves E, Bucher J, Ryll A, Niklas, Mauch K et al. Bridging the layers: Towards integration of signal transduction regulation and metabolism into mathematical models. *Molecular BioSystems* 2013; 9: 1576–83.

33. Milo R, Shen-Orr S, Itzkovitz S, Kashtan N, Chklovskii D, Alon U. Network motifs: Simple building blocks of complex networks. *Science* 2002;298(5594):824–7.

34. Papin JA, Reed JL, Palsson BO. Hierarchical thinking in network biology: The unbiased modularization of biochemical networks. *Trends in Biochemical Sciences* 2004;29(12):641–7.

35. Saez-Rodriguez J, Kremling A, Conzelmann H, Bettenbrock K, Gilles ED. Modular analysis of signal transduction networks. *IEEE Control Systems* 2004;24(4):35–52.

36. Itzkovitz S, Levitt R, Kashtan N, Milo R, Itzkovitz M, Alon U. Coarse-graining and self-dissimilarity of complex networks. *Physical Review E* 2005;71(1):016127.

37. Ingolia NT. Topology and robustness in the Drosophila segment polarity network. *PLoS Biology* 2004;2(6):e123.

38. Ma W, Lai L, Ouyang Q, Tang C. Robustness and modular design of the Drosophila segment polarity network. *Molecular Systems Biology* 2006;2:70.

39. Borisuk MT, Tyson JJ. Bifurcation analysis of a model of mitotic control in frog eggs. *Journal of Theoretical Biology* 1998;195(1):69–85.

40. Hong CI, Conrad ED, Tyson JJ. A proposal for robust temperature compensation of circadian rhythms. *Proceedings of the National Academy of Sciences of the United States of America* 2007;104(4):1195–200.

41. Bornholdt S. Boolean network models of cellular regulation: Prospects and limitations. *Journal of the Royal Society Interface* 2008;5(Suppl 1):S85–94.

42. Davidich M, Bornholdt S. The transition from differential equations to Boolean networks: A case study in simplifying a regulatory network model. *Journal of Theoretical Biology* 2008;255(3):269–77.

43. Norrell J, Samuelsson B, Socolar JE. Attractors in continuous and Boolean networks. *Physical Review E, Statistical, Nonlinear, and Soft Matter Physics* 2007;76(4 Pt 2):046122.

44. Angeli D, Sontag E, eds. *Conference on Mathematical Control Theory 2003*, Baton Rouge, LA, 2003.

45. Enciso GA, Sontag ED. Monotone systems under positive feedback: Multistability and a reduction theorem. *Systems & Control Letters* 2005;54(2):159–68.

46. Chaturvedi R, Huang C, Kazmierczak B, Schneider T, Izaguirre JA, Glimm T et al. On multiscale approaches to three-dimensional modelling of morphogenesis. *Journal of the Royal Society Interface* 2005;2(3):237–53.

47. Newman TJ. Grid-free models of multicellular systems, with an application to large-scale vortices accompanying primitive streak formation. *Current Topics in Developmental Biology* 2008;81:157–82.

48. Schnell S, Maini PK, Newman SA, Newman TJ. Introduction. In *Current Topics in Developmental Biology*, ed. Schatten G, Schnell S, Maini P, Newman SA, Newman T. Vol. 81. Amsterdam: Academic Press; 2008, pp. xvii–xxv.

49. Ciliberto A, Capuani F, Tyson JJ. Modeling networks of coupled enzymatic reactions using the total quasi-steady state approximation. *PLoS Computational Biology* 2007;3(3):e45.

50. Clewley R, Rotstein HG, Kopell N. A computational tool for the reduction of nonlinear ODE systems possessing multiple scales. *Multiscale Modeling & Simulation* 2005;4(3):732–59.

51. Clewley R, Soto-Trevino C, Nadim F. Dominant ionic mechanisms explored in spiking and bursting using local low-dimensional reductions of a biophysically realistic model neuron. *Journal of Computational Neuroscience* 2009;26(1):75–90.

52. Kuwahara H, Myers C, Samoilov M, Barker N, Arkin A. Automated abstraction methodology for genetic regulatory networks. In *Transactions on Computational Systems Biology VI*, ed. Priami C, Plotkin G. Lecture Notes in Computer Science 4220. Berlin: Springer; 2006, pp. 150–75.

53. Jiang P, Ventura AC, Sontag ED, Merajver SD, Ninfa AJ, Del Vecchio D. Load-induced modulation of signal transduction networks. *Science Signaling* 2011;4(194):ra67.

54. Ossareh HR, Ventura AC, Merajver SD, Del Vecchio D. Long signaling cascades tend to attenuate retroactivity. *Biophysical Journal* 2011;100(7):1617–26.

55. Qiao L, Nachbar RB, Kevrekidis IG, Shvartsman SY. Bistability and oscillations in the Huang-Ferrell model of MAPK signaling. *PLoS Computational Biology* 2007;3(9):1819–26.

56. Ventura AC, Jackson TL, Merajver SD. On the role of cell signaling models in cancer research. *Cancer Research* 2009;69(2):400–2.

57. Ventura AC, Jiang P, Van Wassenhove L, Del Vecchio D, Merajver SD, Ninfa AJ. Signaling properties of a covalent modification cycle are altered by a downstream target. *Proceedings of the National Academy of Sciences of the United States of America* 2010;107(22):10032–7.

58. Ventura AC, Sepulchre JA, Merajver SD. A hidden feedback in signaling cascades is revealed. *PLoS Computational Biology* 2008;4(3):e1000041.

59. Wynn ML, Ventura AC, Sepulchre JA, Garcia HJ, Merajver SD. Kinase inhibitors can produce off-target effects and activate linked pathways by retroactivity. *BMC Systems Biology* 2011;5:156.

60. Komarova NL, Zou X, Nie Q, Bardwell L. A theoretical framework for specificity in cell signaling. *Molecular Systems Biology* 2005;1:2005.0023.

61. Müller R. Crosstalk of oncogenic and prostanoid signaling pathways. *Journal of Cancer Research and Clinical Oncology* 2004;130(8):429–44.

62. Tyson JJ, Chen KC, Novak B. Sniffers, buzzers, toggles and blinkers: Dynamics of regulatory and signaling pathways in the cell. *Current Opinion in Cell Biology* 2003;15(2):221–31.

63. Goldbeter A, Koshland DE Jr. An amplified sensitivity arising from covalent modification in biological systems. *Proceedings of the National Academy of Sciences of the United States of America* 1981;78(11):6840–4.

64. Hopfield JJ. Kinetic proofreading: A new mechanism for reducing errors in biosynthetic processes requiring high specificity. *Proceedings of the National Academy of Sciences of the United States of America* 1974;71(10):4135–9.

65. Ma W, Trusina A, El-Samad H, Lim WA, Tang C. Defining network topologies that can achieve biochemical adaptation. *Cell* 2009;138(4):760–73.

66. Shah NA, Sarkar CA. Robust network topologies for generating switch-like cellular responses. *PLoS Computational Biology* 7(6):e1002085.

67. Castillo-Hair SM, Villota ER, Coronado AM. Design principles for robust oscillatory behavior. *Systems and Synthetic Biology* 2015;9(3):125–33.

68. Yan L, Ouyang Q, Wang H. Dose-response aligned circuits in signaling systems. *PLoS One* 2012;7(4):e34727.

69. Chau AH, Walter JM, Gerardin J, Tang C, Lim WA. Designing synthetic regulatory networks capable of self-organizing cell polarization. *Cell* 2012;151(2):320–32.

70. Wang Y, Ku CJ, Zhang ER, Artyukhin AB, Weiner OD, Wu LF et al. Identifying network motifs that buffer front-to-back signaling in polarized neutrophils. *Cell Reports* 2013;3(5):1607–16.

71. Cournac A, Sepulchre JA. Simple molecular networks that respond optimally to time-periodic stimulation. *BMC Systems Biology* 2009;3:29.

72. Yi T-M, Huang Y, Simon MI, Doyle J. Robust perfect adaptation in bacterial chemotaxis through integral feedback control. *Proceedings of the National Academy of Sciences of the United States of America* 2000;97(9):4649–53.

73. Mangan S, Alon U. Structure and function of the feed-forward loop network motif. *Proceedings of the National Academy of Sciences of the United States of America* 2003;100(21):11980–5.

74. Mangan S, Zaslaver A, Alon U. The coherent feedforward loop serves as a sign-sensitive delay element in transcription networks. *Journal of Molecular Biology* 2003;334(2):197–204.

75. Kim D, Kwon YK, Cho KH. The biphasic behavior of incoherent feed-forward loops in biomolecular regulatory networks. *BioEssays: News and Reviews in Molecular, Cellular and Developmental Biology* 2008;30(11–12):1204–11.

76. Khalil HK. *Nonlinear Systems.* 3rd ed. Upper Saddle River, NJ: Prentice Hall; 2002.

77. Kokotovic PV, Khalil HK, O'Reilly J. *Singular Perturbation Methods in Control: Analysis and Design.* Amsterdam: Academic Press; 1986.

78. Kuznetsov YA. *Elements of Applied Bifurcation Theory.* 3rd ed. New York: Springer-Verlag; 2004.

79. Franklin GF, Powell JD, Emani-Naeini A. *Feedback Control of Dynamic Systems.* 5th ed. Upper Saddle River, NJ: Pearson Prentice Hall; 2005.

80. Bentele M, Lavrik I, Ulrich M, Stosser S, Heermann DW, Kalthoff H et al. Mathematical modeling reveals threshold mechanism in CD95-induced apoptosis. *Journal of Cell Biology* 2004;166(6):839–51.

81. Kholodenko BN, Hoek JB, Westerhoff HV, Brown GC. Quantification of information transfer via cellular signal transduction pathways. *FEBS Letters* 1997;414(2):430–4.

82. Kinzer-Ursem TL, Linderman JJ. Both ligand- and cell-specific parameters control ligand agonism in a kinetic model of G protein-coupled receptor signaling. *PLoS Computational Biology* 2007;3(1):e6.

83. Ortega F, Ehrenberg M, Acerenza L, Westerhoff HV, Mas F, Cascante M. Sensitivity analysis of metabolic cascades catalyzed by bifunctional enzymes. *Molecular Biology Reports* 2002;29(1–2):211–5.

84. Sahle S, Mendes P, Hoops S, Kummer U. A new strategy for assessing sensitivities in biochemical models. *Philosophical Transactions Series A, Mathematical, Physical, and Engineering Sciences* 2008;366(1880):3619–31.

85. Wang Z, Deisboeck TS, Cristini V. Development of a sampling-based global sensitivity analysis workflow for multiscale computational cancer models. *IET Systems Biology* 2014;8(5):191–7.

86. Sontag ED. *Mathematical Control Theory: Deterministic Finite Dimensional Systems*. 2nd ed. New York: Springer-Verlag; 1998.

87. Anderson AR, Quaranta V. Integrative mathematical oncology. *Nature Reviews Cancer*. 2008;8(3):227–34.

88. Cristini V, Lowengrub J. *Multiscale Modeling of Cancer: An Integrated Experimental and Mathematical Modeling Approach*. Cambridge: Cambridge University Press; 2010.

89. Galle J, Hoffmann M, Aust G. From single cells to tissue architecture—A bottom-up approach to modelling the spatio-temporal organisation of complex multi-cellular systems. *Journal of Mathematical Biology* 2009;58(1–2):261–83.

90. Wang Z, Deisboeck TS. Computational modeling of brain tumors: Discrete, continuum or hybrid? *Scientific Modeling and Simulation* 2008;15(1–3):381–93.

91. Zhang L, Wang Z, Sagotsky JA, Deisboeck TS. Multiscale agent-based cancer modeling. *Journal of Mathematical Biology* 2009;58(4–5):545–59.

92. Wang Z, Butner JD, Cristini V, Deisboeck TS. Integrated PK-PD and agent-based modeling in oncology. *Journal of Pharmacokinetics and Pharmacodynamics* 2015;42(2):179–89.

93. Davidsson P. Agent based social simulation: A computer science view. *Journal of Artificial Societies and Social Simulation* 2002;5(1).

94. Grimm V. Ten years of individual-based modelling in ecology: What have we learned and what could we learn in the future? *Ecological Modelling* 1999;115(2–3):129–48.

95. Bonabeau E. Agent-based modeling: Methods and techniques for simulating human systems. *Proceedings of the National Academy of Sciences of the United States of America* 2002;99(Suppl 3):7280–7.

96. Beyer T, Meyer-Hermann M. Modeling emergent tissue organization involving high-speed migrating cells in a flow equilibrium. *Physical Review E, Statistical, Nonlinear, and Soft Matter Physics* 2007;76(2 Pt 1):021929.

97. Deisboeck TS, Wang Z, Macklin P, Cristini V. Multiscale cancer modeling. *Annual Review of Biomedical Engineering* 2011;13:127–55.

98. Lowengrub JS, Frieboes HB, Jin F, Chuang YL, Li X, Macklin P et al. Nonlinear modelling of cancer: Bridging the gap between cells and tumours. *Nonlinearity* 2010;23(1):R1–9.

99. Kitano H. Systems biology: A brief overview. *Science* 2002;295(5560):1662–4.

100. Meads MB, Gatenby RA, Dalton WS. Environment-mediated drug resistance: A major contributor to minimal residual disease. *Nature Reviews Cancer* 2009;9(9):665–74.

101. Estes DA, Lovato DM, Khawaja HM, Winter SS, Larson RS. Genetic alterations determine chemotherapy resistance in childhood T-ALL: Modelling in stage-specific cell lines and correlation with diagnostic patient samples. *British Journal of Haematology* 2007;139(1):20–30.

102. Raguz S, Yague E. Resistance to chemotherapy: New treatments and novel insights into an old problem. *British Journal of Cancer* 2008;99(3):387–91.

103. Shibata T, Kokubu A, Gotoh M, Ojima H, Ohta T, Yamamoto M et al. Genetic alteration of Keap1 confers constitutive Nrf2 activation and resistance to chemotherapy in gallbladder cancer. *Gastroenterology* 2008;135(4):1358–68, 68.e1–4.

104. Foo J, Michor F. Evolution of resistance to targeted anti-cancer therapies during continuous and pulsed administration strategies. *PLoS Computational Biology* 2009;5(11):e1000557.

105. Foo J, Michor F. Evolution of resistance to anti-cancer therapy during general dosing schedules. *Journal of Theoretical Biology* 2010;263(2):179–88.

106. Enderling H, Chaplain MA, Anderson AR, Vaidya JS. A mathematical model of breast cancer development, local treatment and recurrence. *Journal of Theoretical Biology* 2007;246(2):245–59.

107. Atari MI, Chappell MJ, Errington RJ, Smith PJ, Evans ND. Kinetic modelling of the role of the aldehyde dehydrogenase enzyme and the breast cancer resistance protein in drug resistance and transport. *Computer Methods and Programs in Biomedicine* 2011;104(2):93–103.

108. Roe-Dale R, Isaacson D, Kupferschmid M. A mathematical model of breast cancer treatment with CMF and doxorubicin. *Bulletin of Mathematical Biology* 2011;73(3):585–608.

109. Bonadonna G, Zambetti M, Valagussa P. Sequential or alternating doxorubicin and CMF regimens in breast cancer with more than three positive nodes. Ten-year results. *JAMA* 1995;273(7):542–7.

110. Justus CR, Dong L, Yang LV. Acidic tumor microenvironment and pH-sensing G protein-coupled receptors. *Frontiers in Physiology* 2013;4:354.

111. Lee HO, Silva AS, Concilio S, Li YS, Slifker M, Gatenby RA et al. Evolution of tumor invasiveness: The adaptive tumor microenvironment landscape model. *Cancer Research* 2011; 71(20):6327–37.

112. Silva AS, Gatenby RA. A theoretical quantitative model for evolution of cancer chemotherapy resistance. *Biology Direct* 2010;5:25.

113. Frieboes HB, Edgerton ME, Fruehauf JP, Rose FR, Worrall LK, Gatenby RA et al. Prediction of drug response in breast cancer using integrative experimental/computational modeling. *Cancer Research* 2009;69(10):4484–92.

114. Gatenby RA, Smallbone K, Maini PK, Rose F, Averill J, Nagle RB et al. Cellular adaptations to hypoxia and acidosis during somatic evolution of breast cancer. *British Journal of Cancer* 2007;97(5):646–53.

115. Smallbone K, Gatenby RA, Gillies RJ, Maini PK, Gavaghan DJ. Metabolic changes during carcinogenesis: Potential impact on invasiveness. *Journal of Theoretical Biology* 2007; 244(4):703–13.

116. Jain RK, Stylianopoulos T. Delivering nanomedicine to solid tumors. *Nature Reviews Clinical Oncology* 2010;7(11):653–64.

117. Swartz MA, Kaipainen A, Netti PA, Brekken C, Boucher Y, Grodzinsky AJ et al. Mechanics of interstitial-lymphatic fluid transport: Theoretical foundation and experimental validation. *Journal of Biomechanics* 1999;32(12):1297–307.

118. Stylianopoulos T, Diop-Frimpong B, Munn LL, Jain RK. Diffusion anisotropy in collagen gels and tumors: The effect of fiber network orientation. *Biophysical Journal* 2010;99(10):3119–28.

119. Stylianopoulos T, Poh MZ, Insin N, Bawendi MG, Fukumura D, Munn LL et al. Diffusion of particles in the extracellular matrix: The effect of repulsive electrostatic interactions. *Biophysical Journal* 2010;99(5):1342–9.

120. Fukumura D, Jain RK. Tumor microvasculature and microenvironment: Targets for anti-angiogenesis and normalization. *Microvascular Research* 2007;74(2–3):72–84.

121. Vakoc BJ, Lanning RM, Tyrrell JA, Padera TP, Bartlett LA, Stylianopoulos T et al. Three-dimensional microscopy of the tumor microenvironment in vivo using optical frequency domain imaging. *Nature Medicine* 2009;15(10):1219–23.

122. Frieboes HB, Jin F, Chuang YL, Wise SM, Lowengrub JS, Cristini V. Three-dimensional multispecies nonlinear tumor growth-II: Tumor invasion and angiogenesis. *Journal of Theoretical Biology* 2010;264(4):1254–78.

123. Macklin P, McDougall S, Anderson AR, Chaplain MA, Cristini V, Lowengrub J. Multiscale modelling and nonlinear simulation of vascular tumour growth. *Journal of Mathematical Biology* 2009;58(4–5):765–98.

124. Wu M, Frieboes HB, McDougall SR, Chaplain MA, Cristini V, Lowengrub J. The effect of interstitial pressure on tumor growth: Coupling with the blood and lymphatic vascular systems. *Journal of Theoretical Biology* 2013;320:131–51.

125. Wu M, Frieboes HB, Chaplain MA, McDougall SR, Cristini V, Lowengrub JS. The effect of interstitial pressure on therapeutic agent transport: Coupling with the tumor blood and lymphatic vascular systems. *Journal of Theoretical Biology* 2014;355:194–207.

126. Jain RK, Tong RT, Munn LL. Effect of vascular normalization by antiangiogenic therapy on interstitial hypertension, peritumor edema, and lymphatic metastasis: Insights from a mathematical model. *Cancer Research* 2007;67(6):2729–35.

127. Wang Z, Butner JD, Kerketta R, Cristini V, Deisboeck TS. Simulating cancer growth with multiscale agent-based modeling. *Seminars in Cancer Biology* 2015;30:70–8.

128. Byrne HM. Dissecting cancer through mathematics: From the cell to the animal model. *Nature Reviews Cancer* 2010;10(3):221–30.

129. Tracqui P. Biophysical models of tumour growth. *Reports on Progress in Physics* 2009; 72(5):056701.

130. Edelman LB, Eddy JA, Price ND. In silico models of cancer. *Wiley Interdisciplinary Reviews Systems Biology and Medicine* 2010;2(4):438–59.

131. Powathil GG, Gordon KE, Hill LA, Chaplain MA. Modelling the effects of cell-cycle heterogeneity on the response of a solid tumour to chemotherapy: Biological insights from a hybrid multiscale cellular automaton model. *Journal of Theoretical Biology* 2012;308:1–19.

132. Billy F, Ribba B, Saut O, Morre-Trouilhet H, Colin T, Bresch D et al. A pharmacologically based multiscale mathematical model of angiogenesis and its use in investigating the efficacy of a new cancer treatment strategy. *Journal of Theoretical Biology* 2009;260(4):545–62.

133. Powathil GG, Swat M, Chaplain MA. Systems oncology: Towards patient-specific treatment regimes informed by multiscale mathematical modelling. *Seminars in Cancer Biology* 2015;30:13–20.

134. Enderling H, Chaplain MA. Mathematical modeling of tumor growth and treatment. *Current Pharmaceutical Design* 2014;20(30):4934–40.

135. Powathil GG, Adamson DJ, Chaplain MA. Towards predicting the response of a solid tumour to chemotherapy and radiotherapy treatments: Clinical insights from a computational model. *PLoS Computational Biology* 2013;9(7):e1003120.

136. Wang Z, Bordas V, Deisboeck TS. Discovering molecular targets in cancer with multiscale modeling. *Drug Development Research* 2011;72(1):45–52.

137. Wang Z, Deisboeck TS. Mathematical modeling in cancer drug discovery. *Drug Discovery Today* 2014;19(2):145–50.

138. Chmielecki J, Foo J, Oxnard GR, Hutchinson K, Ohashi K, Somwar R et al. Optimization of dosing for EGFR-mutant non-small cell lung cancer with evolutionary cancer modeling. *Science Translational Medicine* 2011;3(90):90ra59.

139. Weis JA, Miga MI, Arlinghaus LR, Li X, Abramson V, Chakravarthy AB et al. Predicting the response of breast cancer to neoadjuvant therapy using a mechanically coupled reaction-diffusion model. *Cancer Research* 2015;75(22):4697–707.

140. Weis JA, Miga MI, Arlinghaus LR, Li X, Chakravarthy AB, Abramson V et al. A mechanically coupled reaction-diffusion model for predicting the response of breast tumors to neoadjuvant chemotherapy. *Physics in Medicine and Biology* 2013;58(17):5851–66.

141. Thurber GM, Yang KS, Reiner T, Kohler RH, Sorger P, Mitchison T et al. Single-cell and subcellular pharmacokinetic imaging allows insight into drug action in vivo. *Nature Communications* 2013;4:1504.

142. Venkatasubramanian R, Arenas RB, Henson MA, Forbes NS. Mechanistic modelling of dynamic MRI data predicts that tumour heterogeneity decreases therapeutic response. *British Journal of Cancer* 2010;103(4):486–97.

143. Lee JJ, Huang J, England CG, McNally LR, Frieboes HB. Predictive modeling of in vivo response to gemcitabine in pancreatic cancer. *PLoS Computational Biology* 2013;9(9):e1003231.
144. Gatenby RA, Silva AS, Gillies RJ, Frieden BR. Adaptive therapy. *Cancer Research* 2009; 69(11):4894–903.
145. Hawkins-Daarud A, Rockne R, Corwin D, Anderson ARA, Kinahan P, Swanson KR. In silico analysis suggests differential response to bevacizumab and radiation combination therapy in newly diagnosed glioblastoma. *Journal of the Royal Society Interface* 2015;12(109).
146. Michor F, Hughes TP, Iwasa Y, Branford S, Shah NP, Sawyers CL et al. Dynamics of chronic myeloid leukaemia. *Nature* 2005;435(7046):1267–70.
147. Leder K, Pitter K, Laplant Q, Hambardzumyan D, Ross BD, Chan TA et al. Mathematical modeling of PDGF-driven glioblastoma reveals optimized radiation dosing schedules. *Cell* 2014;156(3):603–16.
148. Stiehl T, Baran N, Ho AD, Marciniak-Czochra A. Cell division patterns in acute myeloid leukemia stem-like cells determine clinical course: A model to predict patient survival. *Cancer Research* 2015;75(6):940–9.
149. Ashley CE, Carnes EC, Phillips GK, Padilla D, Durfee PN, Brown PA et al. The targeted delivery of multicomponent cargos to cancer cells by nanoporous particle-supported lipid bilayers. *Nature Materials* 2011;10(5):389–97.
150. Stapleton S, Milosevic M, Allen C, Zheng J, Dunne M, Yeung I et al. A mathematical model of the enhanced permeability and retention effect for liposome transport in solid tumors. *PLoS One* 2013;8(12):e81157.
151. Lee T-R, Choi M, Kopacz AM, Yun S-H, Liu WK, Decuzzi P. On the near-wall accumulation of injectable particles in the microcirculation: Smaller is not better. *Scientific Reports* 2013;3:2079.
152. van de Ven AL, Wu M, Lowengrub J, McDougall SR, Chaplain MA, Cristini V et al. Integrated intravital microscopy and mathematical modeling to optimize nanotherapeutics delivery to tumors. *AIP Advances* 2012;2(1):11208.
153. Frieboes HB, Wu M, Lowengrub J, Decuzzi P, Cristini V. A computational model for predicting nanoparticle accumulation in tumor vasculature. *PLoS One* 2013;8(2):e56876.
154. Wang Z, Kerketta R, Chuang YL, Dogra P, Butner JD, Brocato TA et al. Theory and experimental validation of a spatio-temporal model of chemotherapy transport to enhance tumor cell kill. *PLoS Computational Biology* 2016;12(6):e1004969.
155. Kim B, Han G, Toley BJ, Kim CK, Rotello VM, Forbes NS. Tuning payload delivery in tumour cylindroids using gold nanoparticles. *Nature Nanotechnology* 2010;5(6):465–72.
156. Enriquez-Navas PM, Kam Y, Das T, Hassan S, Silva A, Foroutan P et al. Exploiting evolutionary principles to prolong tumor control in preclinical models of breast cancer. *Science Translational Medicine* 2016;8(327):327ra24.
157. Frieboes HB, Smith BR, Chuang YL, Ito K, Roettgers AM, Gambhir SS et al. An integrated computational/experimental model of lymphoma growth. *PLoS Computational Biology* 2013;9(3):e1003008.
158. Frieboes HB, Smith BR, Wang Z, Kotsuma M, Ito K, Day A et al. Predictive modeling of drug response in non-Hodgkin's lymphoma. *PLoS One* 2015;10(6):e0129433.
159. Brocato T, Dogra P, Koay EJ, Day A, Chuang YL, Wang Z et al. Understanding drug resistance in breast cancer with mathematical oncology. *Current Breast Cancer Reports* 2014;6(2): 110–20.
160. Merriam-Webster. Pathology. 2015. Available from http://www.merriam-webster.com/dictionary/pathology.
161. Sanders ME, Schuyler PA, Dupont WD, Page DL. The natural history of low-grade ductal carcinoma in situ of the breast in women treated by biopsy only revealed over 30 years of long-term follow-up. *Cancer* 2005;103(12):2481–4.

162. Talsma AK, Reedijk AM, Damhuis RA, Westenend PJ, Vles WJ. Re-resection rates after breast-conserving surgery as a performance indicator: Introduction of a case-mix model to allow comparison between Dutch hospitals. *European Journal of Surgical Oncology: The Journal of the European Society of Surgical Oncology and the British Association of Surgical Oncology* 2011;37(4):357–63.

163. Eccles SA, Aboagye EO, Ali S, Anderson AS, Armes J, Berditchevski F et al. Critical research gaps and translational priorities for the successful prevention and treatment of breast cancer. *Breast Cancer Research: BCR* 2013;15(5):R92.

164. Venkatesan A, Chu P, Kerlikowske K, Sickles EA, Smith-Bindman R. Positive predictive value of specific mammographic findings according to reader and patient variables. *Radiology* 2009;250(3):648–57.

165. Winchester DP, Cox JD. Standards for diagnosis and management of invasive breast carcinoma. American College of Radiology. American College of Surgeons. College of American Pathologists. Society of Surgical Oncology. *CA: A Cancer Journal for Clinicians* 1998; 48(2):83–107.

166. Cabioglu N, Hunt KK, Sahin AA, Kuerer HM, Babiera GV, Singletary SE et al. Role for intraoperative margin assessment in patients undergoing breast-conserving surgery. *Annals of Surgical Oncology* 2007;14(4):1458–71.

167. Cheng L, Al-Kaisi NK, Gordon NH, Liu AY, Gebrail F, Shenk RR. Relationship between the size and margin status of ductal carcinoma in situ of the breast and residual disease. *Journal of the National Cancer Institute* 1997;89(18):1356–60.

168. Dillon MF, Maguire AA, McDermott EW, Myers C, Hill AD, O'Doherty A et al. Needle core biopsy characteristics identify patients at risk of compromised margins in breast conservation surgery. *Modern Pathology: An Official Journal of the United States and Canadian Academy of Pathology, Inc.* 2008;21(1):39–45.

169. Jain RK. Normalizing tumor microenvironment to treat cancer: Bench to bedside to biomarkers. *Journal of Clinical Oncology: Official Journal of the American Society of Clinical Oncology* 2013;31(17):2205–18.

170. Chauhan VP, Stylianopoulos T, Boucher Y, Jain RK. Delivery of molecular and nanoscale medicine to tumors: Transport barriers and strategies. *Annual Review of Chemical and Biomolecular Engineering* 2011;2:281–98.

171. Bankhead A 3rd, Magnuson NS, Heckendorn RB. Cellular automaton simulation examining progenitor hierarchy structure effects on mammary ductal carcinoma in situ. *Journal of Theoretical Biology* 2007;246(3):491–8.

172. Franks SJ, Byrne HM, Mudhar HS, Underwood JC, Lewis CE. Mathematical modelling of comedo ductal carcinoma in situ of the breast. *Mathematical Medicine and Biology: A Journal of the IMA* 2003;20(3):277–308.

173. Franks SJ, Byrne HM, Underwood JC, Lewis CE. Biological inferences from a mathematical model of comedo ductal carcinoma in situ of the breast. *Journal of Theoretical Biology* 2005;232(4):523–43.

174. Gatenby RA, Gawlinski ET, Gmitro AF, Kaylor B, Gillies RJ. Acid-mediated tumor invasion: A multidisciplinary study. *Cancer Research* 2006;66(10):5216–23.

175. Kim Y, Stolarska MA, Othmer HG. The role of the microenvironment in tumor growth and invasion. *Progress in Biophysics and Molecular Biology* 2011;106(2):353–79.

176. Norton KA, Wininger M, Bhanot G, Ganesan S, Barnard N, Shinbrot T. A 2D mechanistic model of breast ductal carcinoma in situ (DCIS) morphology and progression. *Journal of Theoretical Biology* 2010;263(4):393–406.

177. Rejniak KA, Anderson AR. A computational study of the development of epithelial acin. II. Necessary conditions for structure and lumen stability. *Bulletin of Mathematical Biology* 2008;70(5):1450–79.

178. Silva AS, Gatenby RA, Gillies RJ, Yunes JA. A quantitative theoretical model for the development of malignancy in ductal carcinoma in situ. *Journal of Theoretical Biology* 2010; 262(4):601–13.

179. Cristini V, Lowengrub J, Nie Q. Nonlinear simulation of tumor growth. *Journal of Mathematical Biology* 2003;46(3):191–224.

180. Gandhi A, Holland PA, Knox WF, Potten CS, Bundred NJ. Evidence of significant apoptosis in poorly differentiated ductal carcinoma in situ of the breast. *British Journal of Cancer* 1998;78(6):788–94.

181. Macklin P, Edgerton ME, Thompson AM, Cristini V. Patient-calibrated agent-based modelling of ductal carcinoma in situ (DCIS): From microscopic measurements to macroscopic predictions of clinical progression. *Journal of Theoretical Biology* 2012;301:122–40.

182. Hoehme S, Drasdo D. Biomechanical versus nutrient control: What determines the growth dynamics of mammalian cell populations. *Mathematical Population Studies* 2010;17(3):166–87.

183. De Roos MA, Pijnappel RM, Post WJ, De Vries J, Baas PC, Groote LD. Correlation between imaging and pathology in ductal carcinoma in situ of the breast. *World Journal of Surgical Oncology* 2004;2:4.

184. O'Connor OA, Horwitz S, Hamlin P, Portlock C, Moskowitz CH, Sarasohn D et al. Phase II-I-II study of two different doses and schedules of pralatrexate, a high-affinity substrate for the reduced folate carrier, in patients with relapsed or refractory lymphoma reveals marked activity in T-cell malignancies. *Journal of Clinical Oncology: Official Journal of the American Society of Clinical Oncology* 2009;27(26):4357–64.

185. O'Connor OA, Pro B, Pinter-Brown L, Bartlett N, Popplewell L, Coiffier B et al. Pralatrexate in patients with relapsed or refractory peripheral T-cell lymphoma: Results from the pivotal PROPEL study. *Journal of Clinical Oncology: Official Journal of the American Society of Clinical Oncology* 2011;29(9):1182–9.

186. Toner LE, Vrhovac R, Smith EA, Gardner J, Heaney M, Gonen M et al. The schedule-dependent effects of the novel antifolate pralatrexate and gemcitabine are superior to methotrexate and cytarabine in models of human non-Hodgkin's lymphoma. *Clinical Cancer Research: An Official Journal of the American Association for Cancer Research* 2006;12(3 Pt 1):924–32.

187. Schmitt CA, Fridman JS, Yang M, Lee S, Baranov E, Hoffman RM et al. A senescence program controlled by p53 and p16INK4a contributes to the outcome of cancer therapy. *Cell* 2002;109(3):335–46.

188. Lowe SW, Sherr CJ. Tumor suppression by Ink4a-Arf: Progress and puzzles. *Current Opinion in Genetics & Development* 2003;13(1):77–83.

189. Das H, Wang Z, Niazi MK, Aggarwal R, Lu J, Kanji S et al. Impact of diffusion barriers to small cytotoxic molecules on the efficacy of immunotherapy in breast cancer. *PLoS One* 2013;8(4):e61398.

190. Total cost of cancer care by site of service: Phyician office vs outpatient hospital. Washington, DC: Avalere Health; 2012. Available from http://www.communityoncology.org/pdfs/avalere-cost-of-cancer-care-study.pdf.

191. Raspollini MR, Castiglione F, Rossi Degl'innocenti D, Amunni G, Villanucci A, Garbini F et al. Tumour-infiltrating gamma/delta T-lymphocytes are correlated with a brief disease-free interval in advanced ovarian serous carcinoma. *Annals of Oncology* 2005;16(4):590–6.

192. Cancer patients receiving chemotherapy: Opportunities for better management. New York: Milliman Inc.; 2010. Available from http://www.milliman.com/uploadedFiles/insight/research/health-rr/cancer-patients-receiving-chemotherapy.pdf.

193. Treatment planning. Stages I and II breast cancer. Fort Washington, PA: National Comprehensive Cancer Network; 2014. Available from http://www.nccn.org/patients/guidelines/stage_i_ii_breast/index.html#78.

194. Bird RB, Stewart WE, Lightfoot EN. *Transport Phenomena.* Wiley; 2007.

195. Zill DG, Wright WS, Cullen MR. *Differential Equations with Boundary-Value Problems.* Boston: Brooks Cole Publishing; 2012.

196. Baxter LT, Jain RK. Pharmacokinetic analysis of the microscopic distribution of enzyme-conjugated antibodies and prodrugs: Comparison with experimental data. *British Journal of Cancer* 1996;73(4):447–56.

197. El-Kareh AW, Secomb TW. Two-mechanism peak concentration model for cellular pharmacodynamics of doxorubicin. *Neoplasia* 2005;7(7):705–13.

198. Jackson TL. Intracellular accumulation and mechanism of action of doxorubicin in a spatiotemporal tumor model. *Journal of Theoretical Biology* 2003;220(2):201–13.

199. Levenspiel O. *Chemical Reaction Engineering.* New York: Wiley; 1999.

200. Ashley CE, Carnes EC, Epler KE, Padilla DP, Phillips GK, Castillo RE et al. Delivery of small interfering RNA by peptide-targeted mesoporous silica nanoparticle-supported lipid bilayers. *ACS Nano* 2012;6(3):2174–88.

201. Women's health. Geneva: World Health Organization; cited 2015. Available from http://www.who.int/mediacentre/factsheets/fs334/en/.

202. Rosenberg SA, Yang JC, Restifo NP. Cancer immunotherapy: Moving beyond current vaccines. *Nature Medicine* 2004;10(9):909–15.

203. Bank I, Book M, Huszar M, Baram Y, Schnirer I, Brenner H. V delta 2+ gamma delta T lymphocytes are cytotoxic to the MCF 7 breast carcinoma cell line and can be detected among the T cells that infiltrate breast tumors. *Clinical Immunology and Immunopathology* 1993;67(1):17–24.

204. Brenner MB, McLean J, Dialynas DP, Strominger JL, Smith JA, Owen FL et al. Identification of a putative second T-cell receptor. *Nature* 1986;322(6075):145–9.

205. Shin S, El-Diwany R, Schaffert S, Adams EJ, Garcia KC, Pereira P et al. Antigen recognition determinants of gammadelta T cell receptors. *Science* 2005;308(5719):252–5.

206. Ikeda H, Old LJ, Schreiber RD. The roles of IFN gamma in protection against tumor development and cancer immunoediting. *Cytokine & Growth Factor Reviews* 2002;13(2):95–109.

207. Wang L, Kamath A, Das H, Li L, Bukowski JF. Antibacterial effect of human V gamma 2V delta 2 T cells in vivo. *Journal of Clinical Investigation.* 2001;108(9):1349–57.

208. Das H, Groh V, Kuijl C, Sugita M, Morita CT, Spies T et al. MICA engagement by human Vgamma2Vdelta2 T cells enhances their antigen-dependent effector function. *Immunity* 2001;15(1):83–93.

209. Goel S, Wong AH, Jain RK. Vascular normalization as a therapeutic strategy for malignant and nonmalignant disease. *Cold Spring Harbor Perspectives in Medicine* 2012;2(3):a006486.

210. Huang Y, Goel S, Duda DG, Fukumura D, Jain RK. Vascular normalization as an emerging strategy to enhance cancer immunotherapy. *Cancer Research* 2013;73(10):2943–8.

211. Huang Y, Stylianopoulos T, Duda DG, Fukumura D, Jain RK. Benefits of vascular normalization are dose and time dependent—Letter. *Cancer Research* 2013;73(23):7144–6.

212. Kirui DK, Koay EJ, Guo X, Cristini V, Shen H, Ferrari M. Tumor vascular permeabilization using localized mild hyperthermia to improve macromolecule transport. *Nanomedicine: Nanotechnology, Biology, and Medicine* 2014;10(7):1487–96.

213. Tasseff R, Nayak S, Salim S, Kaushik P, Rizvi N, Varner JD. Analysis of the molecular networks in androgen dependent and independent prostate cancer revealed fragile and robust subsystems. *PLoS One* 2010;5(1):e8864.

214. Wang Z, Birch CM, Sagotsky J, Deisboeck TS. Cross-scale, cross-pathway evaluation using an agent-based non-small cell lung cancer model. *Bioinformatics* 2009;25(18):2389–96.

215. Wang Z, Bordas V, Deisboeck TS. Identification of critical molecular components in a multiscale cancer model based on the integration of Monte Carlo, resampling, and ANOVA. *Frontiers in Physiology* 2011;2:35.

216. Pasquier E, Kavallaris M, Andre N. Metronomic chemotherapy: New rationale for new directions. *Nature Reviews Clinical Oncology* 2010;7(8):455–65.

217. Olive KP, Jacobetz MA, Davidson CJ, Gopinathan A, McIntyre D, Honess D et al. Inhibition of Hedgehog signaling enhances delivery of chemotherapy in a mouse model of pancreatic cancer. *Science* 2009;324(5933):1457–61.

218. Provenzano PP, Cuevas C, Chang AE, Goel VK, Von Hoff DD, Hingorani SR. Enzymatic targeting of the stroma ablates physical barriers to treatment of pancreatic ductal adenocarcinoma. *Cancer Cell* 2012;21(3):418–29.

219. Shen H, Rodriguez-Aguayo C, Xu R, Gonzalez-Villasana V, Mai J, Huang Y et al. Enhancing chemotherapy response with sustained EphA2 silencing using multistage vector delivery. *Clinical Cancer Research: An Official Journal of the American Association for Cancer Research* 2013;19(7):1806–15.

220. Tanaka T, Mangala LS, Vivas-Mejia PE, Nieves-Alicea R, Mann AP, Mora E et al. Sustained small interfering RNA delivery by mesoporous silicon particles. *Cancer Research* 2010;70(9):3687–96.

221. Xu R, Huang Y, Mai J, Zhang G, Guo X, Xia X et al. Multistage vectored siRNA targeting ataxia-telangiectasia mutated for breast cancer therapy. *Small* 2013;9(9–10):1799–808.

222. Dai C-L, Xiong H-Y, Tang L-F, Zhang X, Liang Y-J, Zeng M-S et al. Tetrandrine achieved plasma concentrations capable of reversing MDR in vitro and had no apparent effect on doxorubicin pharmacokinetics in mice. *Cancer Chemotherapy and Pharmacology* 2007; 60(5):741–50.

223. Lu WL, Qi XR, Zhang Q, Li RY, Wang GL, Zhang RJ et al. A pegylated liposomal platform: Pharmacokinetics, pharmacodynamics, and toxicity in mice using doxorubicin as a model drug. *Journal of Pharmacological Sciences* 2004;95(3):381–9.

224. Richly H, Grubert M, Scheulen ME, Hilger RA. Plasma and cellular pharmacokinetics of doxorubicin after intravenous infusion of Caelyx/Doxil in patients with hematological tumors. *International Journal of Clinical Pharmacology and Therapeutics* 2009;47(1):55–7.

225. Decuzzi P, Godin B, Tanaka T, Lee SY, Chiappini C, Liu X et al. Size and shape effects in the biodistribution of intravascularly injected particles. *Journal of Controlled Release: Official Journal of the Controlled Release Society* 2010;141(3):320–7.

226. Junttila MR, de Sauvage FJ. Influence of tumour micro-environment heterogeneity on therapeutic response. *Nature* 2013;501(7467):346–54.

227. Kubitza M, Hickey L, Roberts WG. Influence of host microvascular environment on tumour vascular endothelium. *International Journal of Experimental Pathology* 1999; 80(1):1–10.

228. Gilbert LA, Hemann MT. DNA damage-mediated induction of a chemoresistant niche. *Cell* 2010;143(3):355–66.

229. Kreyszig E. *Advanced Engineering Mathematics.* Hoboken, NJ: John Wiley & Sons; 2010.

230. Stewart J. *Calculus.* Boston: Thomson Brooks/Cole; 2008.

231. Team TG. GNU Image Manipulation Program. Available from https://docs.gimp.org/2.8/en/.

232. Minchinton AI, Tannock IF. Drug penetration in solid tumours. *Nature Reviews Cancer* 2006;6(8):583–92.

233. Tredan O, Galmarini CM, Patel K, Tannock IF. Drug resistance and the solid tumor microenvironment. *Journal of the National Cancer Institute* 2007;99(19):1441–54.

234. Wang Z, Birch CM, Deisboeck TS. Cross-scale sensitivity analysis of a non-small cell lung cancer model: Linking molecular signaling properties to cellular behavior. *Bio Systems* 2008; 92(3):249–58.

235. Wang Z, Bordas V, Sagotsky J, Deisboeck TS. Identifying therapeutic targets in a combined EGFR-TGFbetaR signalling cascade using a multiscale agent-based cancer model. *Mathematical Medicine and Biology: A Journal of the IMA* 2012;29(1):95–108.

236. Lennon AM, Wolfgang CL, Canto MI, Klein AP, Herman JM, Goggins M et al. The early detection of pancreatic cancer: What will it take to diagnose and treat curable pancreatic neoplasia? *Cancer Research* 2014;74(13):3381–9.

237. Evans DB, Multidisciplinary Pancreatic Cancer Study Group. Resectable pancreatic cancer: The role for neoadjuvant/preoperative therapy. *HPB: The Official Journal of the International Hepato Pancreato Biliary Association* 2006;8(5):365–8.

238. Callery MP, Chang KJ, Fishman EK, Talamonti MS, William Traverso L, Linehan DC. Pretreatment assessment of resectable and borderline resectable pancreatic cancer: Expert consensus statement. *Annals of Surgical Oncology* 2009;16(7):1727–33.

239. Varadhachary GR, Tamm EP, Abbruzzese JL, Xiong HQ, Crane CH, Wang H et al. Borderline resectable pancreatic cancer: Definitions, management, and role of preoperative therapy. *Annals of Surgical Oncology* 2006;13(8):1035–46.

240. Katz MH, Pisters PW, Evans DB, Sun CC, Lee JE, Fleming JB et al. Borderline resectable pancreatic cancer: The importance of this emerging stage of disease. *Journal of the American College of Surgeons* 2008;206(5):833–46; discussion 46–8.

241. Crane CH, Varadhachary G, Pisters PWT, Evans DB, Wolff RA. Future chemoradiation strategies in pancreatic cancer. *Seminars in Oncology* 2007;34(4):335–46.

242. Chatterjee D, Katz MH, Rashid A, Varadhachary GR, Wolff RA, Wang H et al. Histologic grading of the extent of residual carcinoma following neoadjuvant chemoradiation in pancreatic ductal adenocarcinoma: A predictor for patient outcome. *Cancer* 2012;118(12):3182–90.

243. Crawley AS, O'Kennedy RJ. The need for effective pancreatic cancer detection and management: A biomarker-based strategy. *Expert Review of Molecular Diagnostics* 2015;15(10):1339–53.

244. Ferrari M. Frontiers in cancer nanomedicine: Directing mass transport through biological barriers. *Trends in Biotechnology* 2010;28(4):181–8.

245. Poplin E, Wasan H, Rolfe L, Raponi M, Ikdahl T, Bondarenko I et al. Randomized, multicenter, phase II study of CO-101 versus gemcitabine in patients with metastatic pancreatic ductal adenocarcinoma: Including a prospective evaluation of the role of hENT1 in gemcitabine or CO-101 sensitivity. *Journal of Clinical Oncology: Official Journal of the American Society of Clinical Oncology* 2013;31(35):4453–61.

246. Itakura J, Ishiwata T, Shen B, Kornmann M, Korc M. Concomitant over–expression of vascular endothelial growth factor and its receptors in pancreatic cancer. *International Journal of Cancer* 2000;85(1):27–34.

247. Adams RH, Alitalo K. Molecular regulation of angiogenesis and lymphangiogenesis. *Nature Reviews Molecular Cell Biology* 2007;8(6):464–78.

248. Kerbel RS. Tumor angiogenesis. *New England Journal of Medicine* 2008;358(19):2039–49.

249. Whipple C, Korc M. Targeting angiogenesis in pancreatic cancer: Rationale and pitfalls. *Langenbeck's Archives of Surgery* 2008;393(6):901–10.

250. Li A, Dubey S, Varney ML, Dave BJ, Singh RK. IL-8 directly enhanced endothelial cell survival, proliferation, and matrix metalloproteinases production and regulated angiogenesis. *Journal of Immunology* 2003;170(6):3369–76.

251. Albini A, Tosetti F, Li VW, Noonan DM, Li WW. Cancer prevention by targeting angiogenesis. *Nature Reviews Clinical Oncology* 2012;9(9):498–509.

252. Crane CH, Winter K, Regine WF, Safran H, Rich TA, Curran W et al. Phase II study of bevacizumab with concurrent capecitabine and radiation followed by maintenance gemcitabine and bevacizumab for locally advanced pancreatic cancer: Radiation Therapy Oncology Group RTOG 0411. *Journal of Clinical Oncology: Official Journal of the American Society of Clinical Oncology* 2009;27(25):4096–102.

253. Kindler HL, Niedzwiecki D, Hollis D, Sutherland S, Schrag D, Hurwitz H et al. Gemcitabine plus bevacizumab compared with gemcitabine plus placebo in patients with advanced pancreatic cancer: Phase III trial of the Cancer and Leukemia Group B (CALGB 80303). *Journal of Clinical Oncology: Official Journal of the American Society of Clinical Oncology* 2010;28(22):3617–22.

254. Chu GC, Kimmelman AC, Hezel AF, DePinho RA. Stromal biology of pancreatic cancer. *Journal of Cellular Biochemistry* 2007;101(4):887–907.

255. Angeli F, Koumakis G, Chen MC, Kumar S, Delinassios JG. Role of stromal fibroblasts in cancer: Promoting or impeding? *Tumour Biology: The Journal of the International Society for Oncodevelopmental Biology and Medicine* 2009;30(3):109–20.

256. Ozdemir BC, Pentcheva-Hoang T, Carstens JL, Zheng X, Wu CC, Simpson TR et al. Depletion of carcinoma-associated fibroblasts and fibrosis induces immunosuppression and accelerates pancreas cancer with reduced survival. *Cancer Cell* 2014;25(6):719–34.

257. Rhim AD, Oberstein PE, Thomas DH, Mirek ET, Palermo CF, Sastra SA et al. Stromal elements act to restrain, rather than support, pancreatic ductal adenocarcinoma. *Cancer Cell* 2014;25(6):735–47.

258. Viale A, Pettazzoni P, Lyssiotis CA, Ying H, Sanchez N, Marchesini M et al. Oncogene ablation-resistant pancreatic cancer cells depend on mitochondrial function. *Nature* 2014; 514(7524):628–32.

259. Thayer SP, di Magliano MP, Heiser PW, Nielsen CM, Roberts DJ, Lauwers GY et al. Hedgehog is an early and late mediator of pancreatic cancer tumorigenesis. *Nature* 2003;425(6960):851–6.

260. Ko AH, LoConte N, Tempero MA, Walker EJ, Kate Kelley R, Lewis S et al. A phase I study of FOLFIRINOX plus IPI-926, a hedgehog pathway inhibitor, for advanced pancreatic adenocarcinoma. *Pancreas* 2016;45(3):370–5.

261. Neesse A, Frese KK, Bapiro TE, Nakagawa T, Sternlicht MD, Seeley TW et al. CTGF antagonism with mAb FG-3019 enhances chemotherapy response without increasing drug delivery in murine ductal pancreas cancer. *Proceedings of the National Academy of Sciences of the United States of America* 2013;110(30):12325–30.

262. Neesse A, Krug S, Gress TM, Tuveson DA, Michl P. Emerging concepts in pancreatic cancer medicine: Targeting the tumor stroma. *OncoTargets and Therapy* 2013;7:33–43.

263. Chen WY, Abatangelo G. Functions of hyaluronan in wound repair. *Wound Repair and Regeneration: Official Publication of the Wound Healing Society [and] the European Tissue Repair Society* 1999;7(2):79–89.

264. Chlenski A, Liu S, Guerrero LJ, Yang Q, Tian Y, Salwen HR et al. SPARC expression is associated with impaired tumor growth, inhibited angiogenesis and changes in the extracellular matrix. *International Journal of Cancer* 2006;118(2):310–6.

265. Podhajcer OL, Benedetti L, Girotti MR, Prada F, Salvatierra E, Llera AS. The role of the matricellular protein SPARC in the dynamic interaction between the tumor and the host. *Cancer Metastasis Reviews* 2008;27(3):523–37.

266. Neesse A, Michl P, Frese KK, Feig C, Cook N, Jacobetz MA et al. Stromal biology and therapy in pancreatic cancer. *Gut* 2011;60(6):861–8.

267. Infante JR, Matsubayashi H, Sato N, Tonascia J, Klein AP, Riall TA et al. Peritumoral fibroblast SPARC expression and patient outcome with resectable pancreatic adenocarcinoma. *Journal of Clinical Oncology* 2007;25(3):319–25.

268. Desai N, Trieu V, Damascelli B, Soon-Shiong P. SPARC expression correlates with tumor response to albumin-bound paclitaxel in head and neck cancer patients. *Translational Oncology* 2009;2(2):59–64.

269. Krusius T, Ruoslahti E. Primary structure of an extracellular matrix proteoglycan core protein deduced from cloned cDNA. *Proceedings of the National Academy of Sciences of the United States of America* 1986;83(20):7683–7.

270. Iozzo RV. The family of the small leucine-rich proteoglycans: Key regulators of matrix assembly and cellular growth. *Critical Reviews in Biochemistry and Molecular Biology* 1997;32(2):141–74.

271. Williams KE, Fulford LA, Albig AR. Lumican reduces tumor growth via induction of Fas-mediated endothelial cell apoptosis. *Cancer Microenvironment* 2011;4(1):115–26.

272. Li X, Truty MA, Kang Y, Chopin-Laly X, Zhang R, Roife D et al. Extracellular lumican inhibits pancreatic cancer cell growth and is associated with prolonged survival after surgery. *Clinical Cancer Research: An Official Journal of the American Association for Cancer Research* 2014;20(24):6529–40.

273. Gapstur SM, Gann PH, Lowe W, Liu K, Colangelo L, Dyer A. Abnormal glucose metabolism and pancreatic cancer mortality. *JAMA* 2000;283(19):2552–8.

274. Son J, Lyssiotis CA, Ying H, Wang X, Hua S, Ligorio M et al. Glutamine supports pancreatic cancer growth through a KRAS-regulated metabolic pathway. *Nature* 2013;496(7443):101–5.

275. Swietach P, Vaughan-Jones RD, Harris AL. Regulation of tumor pH and the role of carbonic anhydrase 9. *Cancer Metastasis Reviews* 2007;26(2):299–310.

276. Fischer K, Hoffmann P, Voelkl S, Meidenbauer N, Ammer J, Edinger M et al. Inhibitory effect of tumor cell-derived lactic acid on human T cells. *Blood* 2007;109(9):3812–9.

277. Lai I-L, Chou C-C, Lai P-T, Fang C-S, Shirley LA, Yan R et al. Targeting the Warburg effect with a novel glucose transporter inhibitor to overcome gemcitabine resistance in pancreatic cancer cells. *Carcinogenesis* 2014;35(10):2203–13.

278. Koido S, Homma S, Takahara A, Namiki Y, Tsukinaga S, Mitobe J et al. Current immuno-therapeutic approaches in pancreatic cancer. *Clinical & Developmental Immunology* 2011; 2011:267539.

279. Clark CE, Beatty GL, Vonderheide RH. Immunosurveillance of pancreatic adenocarci-noma: Insights from genetically engineered mouse models of cancer. *Cancer Letters* 2009; 279(1):1–7.

280. Lunardi S, Muschel RJ, Brunner TB. The stromal compartments in pancreatic cancer: Are there any therapeutic targets? *Cancer Letters* 2014;343(2):147–55.

281. Laheru D, Jaffee EM. Immunotherapy for pancreatic cancer—Science driving clinical prog-ress. *Nature Reviews Cancer* 2005;5(6):459–67.

282. Soares KC, Zheng L, Edil B, Jaffee EM. Vaccines for pancreatic cancer. *Cancer Journal* 2012; 18(6):642–52.

283. Lutz ER, Wu AA, Bigelow E, Sharma R, Mo G, Soares K et al. Immunotherapy converts nonimmunogenic pancreatic tumors into immunogenic foci of immune regulation. *Cancer Immunology Research* 2014;2(7):616–31.

284. Ino Y, Yamazaki-Itoh R, Shimada K, Iwasaki M, Kosuge T, Kanai Y et al. Immune cell infiltra-tion as an indicator of the immune microenvironment of pancreatic cancer. *British Journal of Cancer* 2013;108(4):914–23.

285. Ishida Y, Agata Y, Shibahara K, Honjo T. Induced expression of PD-1, a novel member of the immunoglobulin gene superfamily, upon programmed cell death. *EMBO Journal* 1992; 11(11):3887–95.

286. Nomi T, Sho M, Akahori T, Hamada K, Kubo A, Kanehiro H et al. Clinical significance and therapeutic potential of the programmed death-1 ligand/programmed death-1 pathway in human pancreatic cancer. *Clinical Cancer Research: An Official Journal of the American Association for Cancer Research* 2007;13(7):2151–7.

287. Topalian SL, Hodi FS, Brahmer JR, Gettinger SN, Smith DC, McDermott DF et al. Safety, activity, and immune correlates of anti-PD-1 antibody in cancer. *New England Journal of Medicine* 2012;366(26):2443–54.

288. Jacobetz MA, Chan DS, Neesse A, Bapiro TE, Cook N, Frese KK et al. Hyaluronan impairs vascular function and drug delivery in a mouse model of pancreatic cancer. *Gut* 2013;62(1):112–20.

289. Kobold S, Grassmann S, Chaloupka M, Lampert C, Wenk S, Kraus F et al. Impact of a new fusion receptor on PD-1–mediated immunosuppression in adoptive T cell therapy. *Journal of the National Cancer Institute* 2015;107(8).

290. Wong PP, Demircioglu F, Ghazaly E, Alrawashdeh W, Stratford MR, Scudamore CL et al. Dual-action combination therapy enhances angiogenesis while reducing tumor growth and spread. *Cancer Cell* 2015;27(1):123–37.

291. Bauer C, Bauernfeind F, Sterzik A, Orban M, Schnurr M, Lehr HA et al. Dendritic cell-based vaccination combined with gemcitabine increases survival in a murine pancreatic carcinoma model. *Gut* 2007;56(9):1275–82.

292. Le DT, Wang-Gillam A, Picozzi V, Greten TF, Crocenzi T, Springett G et al. Safety and survival with GVAX pancreas prime and *Listeria monocytogenes*-expressing mesothelin (CRS-207) boost vaccines for metastatic pancreatic cancer. *Journal of Clinical Oncology: Official Journal of the American Society of Clinical Oncology* 2015;33(12):1325–33.

293. Le DT, Uram JN, Wang H, Bartlett BR, Kemberling H, Eyring AD et al. PD-1 blockade in tumors with mismatch-repair deficiency. *New England Journal of Medicine* 2015;372(26):2509–20.

294. Huguet F, Hammel P, Vernerey D, Goldstein D, Laethem JLV, Glimelius B et al. Impact of chemoradiotherapy (CRT) on local control and time without treatment in patients with locally advanced pancreatic cancer (LAPC) included in the international phase III LAP 07 study. *Journal of Clinical Oncology: Official Journal of the American Society of Clinical Oncology* 2014;32(5s):2014 (suppl; abstr 4001).

295. Lou KJ. Stromal uncertainties in pancreatic cancer. SciBX 2014; doi: 10.1038/scibx.2014.665.

296. What are the key statistics about pancreatic cancer? Atlanta, GA: American Cancer Society; 2015. Available from http://www.cancer.org/cancer/pancreaticcancer/detailedguide/pancreatic-cancer-key-statistics.

297. Rahib L, Smith BD, Aizenberg R, Rosenzweig AB, Fleshman JM, Matrisian LM. Projecting cancer incidence and deaths to 2030: The unexpected burden of thyroid, liver, and pancreas cancers in the United States. *Cancer Research* 2014;74(11):2913–21.

298. Crane CH, Iacobuzio-Donahue CA. Keys to personalized care in pancreatic oncology. *Journal of Clinical Oncology: Official Journal of the American Society of Clinical Oncology* 2012;30(33):4049–50.

299. Li D, Xie K, Wolff R, Abbruzzese JL. Pancreatic cancer. *Lancet* 2004;363(9414):1049–57.

300. Maitra A, Hruban RH. Pancreatic cancer. *Annual Review of Pathology* 2008;3:157–88.

301. Beatty GL, Chiorean EG, Fishman MP, Saboury B, Teitelbaum UR, Sun W et al. CD40 agonists alter tumor stroma and show efficacy against pancreatic carcinoma in mice and humans. *Science* 2011;331(6024):1612–6.

302. Pries AR, Hopfner M, le Noble F, Dewhirst MW, Secomb TW. The shunt problem: Control of functional shunting in normal and tumour vasculature. *Nature Reviews Cancer* 2010; 10(8):587–93.

303. Stylianopoulos T, Martin JD, Chauhan VP, Jain SR, Diop-Frimpong B, Bardeesy N et al. Causes, consequences, and remedies for growth-induced solid stress in murine and human tumors. *Proceedings of the National Academy of Sciences of the United States of America* 2012;109(38):15101–8.

304. Fleischmann D, Rubin GD, Bankier AA, Hittmair K. Improved uniformity of aortic enhancement with customized contrast medium injection protocols at CT angiography. *Radiology* 2000;214(2):363–71.

305. Heiken JP, Brink JA, McClennan BL, Sagel SS, Forman HP, DiCroce J. Dynamic contrast-enhanced CT of the liver: Comparison of contrast medium injection rates and uniphasic and biphasic injection protocols. *Radiology* 1993;187(2):327–31.

306. Nakayama Y, Awai K, Yanaga Y, Nakaura T, Funama Y, Hirai T et al. Optimal contrast medium injection protocols for the depiction of the Adamkiewicz artery using 64-detector CT angiography. *Clinical Radiology* 2008;63(8):880–7.

307. Koay EJ, Baio FE, Ondari A, Truty MJ, Cristini V, Thomas RM et al. Intra-tumoral heterogeneity of gemcitabine delivery and mass transport in human pancreatic cancer. *Physical Biology* 2014;11(6):065002.

308. Farrell JJ, Elsaleh H, Garcia M, Lai R, Ammar A, Regine WF et al. Human equilibrative nucleoside transporter 1 levels predict response to gemcitabine in patients with pancreatic cancer. *Gastroenterology* 2009;136(1):187–95.

309. Zhao Q, Rashid A, Gong Y, Katz MH, Lee JE, Wolf R et al. Pathologic complete response to neoadjuvant therapy in patients with pancreatic ductal adenocarcinoma is associated with a better prognosis. *Annals of Diagnostic Pathology* 2012;16(1):29–37.

310. Evans DB, Varadhachary GR, Crane CH, Sun CC, Lee JE, Pisters PW et al. Preoperative gemcitabine-based chemoradiation for patients with resectable adenocarcinoma of the pancreatic head. *Journal of Clinical Oncology: Official Journal of the American Society of Clinical Oncology* 2008;26(21):3496–502.

311. Varadhachary GR, Wolff RA, Crane CH, Sun CC, Lee JE, Pisters PW et al. Preoperative gemcitabine and cisplatin followed by gemcitabine-based chemoradiation for resectable adenocarcinoma of the pancreatic head. *Journal of Clinical Oncology: Official Journal of the American Society of Clinical Oncology* 2008;26(21):3487–95.

312. Eisbruch A, Shewach DS, Bradford CR, Littles JF, Teknos TN, Chepeha DB et al. Radiation concurrent with gemcitabine for locally advanced head and neck cancer: A phase I trial and intracellular drug incorporation study. *Journal of Clinical Oncology: Official Journal of the American Society of Clinical Oncology* 2001;19(3):792–9.

313. Taghian AG, Abi-Raad R, Assaad SI, Casty A, Ancukiewicz M, Yeh E et al. Paclitaxel decreases the interstitial fluid pressure and improves oxygenation in breast cancers in patients treated with neoadjuvant chemotherapy: Clinical implications. *Journal of Clinical Oncology: Official Journal of the American Society of Clinical Oncology* 2005;23(9):1951–61.

314. Koay EJ, Lee Y, Cristini V, Lowengrub J, Kang Y, Almahariq M et al. A visually apparent and quantifiable CT imaging feature identifies biophysical subtypes of pancreatic ductal adenocarcinoma. 2016; submitted.

315. Hidalgo M. Pancreatic cancer. *New England Journal of Medicine* 2010;362(17):1605–17.

316. Bever KM, Sugar EA, Bigelow E, Sharma R, Laheru D, Wolfgang CL et al. The prognostic value of stroma in pancreatic cancer in patients receiving adjuvant therapy. *HPB: The Official Journal of the International Hepato Pancreato Biliary Association* 2015;17(4):292–8.

317. Hwang RF, Moore T, Arumugam T, Ramachandran V, Amos KD, Rivera A et al. Cancer-associated stromal fibroblasts promote pancreatic tumor progression. *Cancer Research* 2008; 68(3):918–26.

318. Bearer EL, Lowengrub JS, Frieboes HB, Chuang YL, Jin F, Wise SM et al. Multiparameter computational modeling of tumor invasion. *Cancer Research* 2009;69(10):4493–501.

319. Cristini V, Frieboes HB, Gatenby R, Caserta S, Ferrari M, Sinek J. Morphologic instability and cancer invasion. *Clinical Cancer Research: An Official Journal of the American Association for Cancer Research* 2005;11(19 Pt 1):6772–9.

320. Cristini V, Li X, Lowengrub JS, Wise SM. Nonlinear simulations of solid tumor growth using a mixture model: Invasion and branching. *Journal of Mathematical Biology* 2009; 58(4–5):723–63.

321. Frieboes HB, Lowengrub JS, Wise S, Zheng X, Macklin P, Bearer EL et al. Computer simulation of glioma growth and morphology. *NeuroImage* 2007;37(Suppl 1):S59–70.

322. Collisson EA, Sadanandam A, Olson P, Gibb WJ, Truitt M, Gu S et al. Subtypes of pancreatic ductal adenocarcinoma and their differing responses to therapy. *Nature Medicine* 2011;17(4):500–3.

323. Moffitt RA, Marayati R, Flate EL, Volmar KE, Loeza SG, Hoadley KA et al. Virtual microdissection identifies distinct tumor- and stroma-specific subtypes of pancreatic ductal adenocarcinoma. *Nature Genetics* 2015;47(10):1168–78.

324. Gajewski TF, Schreiber H, Fu YX. Innate and adaptive immune cells in the tumor microenvironment. *Nature Immunology* 2013;14(10):1014–22.

325. Salama P, Phillips M, Grieu F, Morris M, Zeps N, Joseph D et al. Tumor-infiltrating FOXP3+ T regulatory cells show strong prognostic significance in colorectal cancer. *Journal of Clinical Oncology: Official Journal of the American Society of Clinical Oncology* 2009;27(2):186–92.

326. de Jong RA, Leffers N, Boezen HM, ten Hoor KA, van der Zee AG, Hollema H et al. Presence of tumor-infiltrating lymphocytes is an independent prognostic factor in type I and II endometrial cancer. *Gynecologic Oncology* 2009;114(1):105–10.

327. Menon AG, Janssen-van Rhijn CM, Morreau H, Putter H, Tollenaar RA, van de Velde CJ et al. Immune system and prognosis in colorectal cancer: A detailed immunohistochemical analysis. *Laboratory Investigation; a Journal of Technical Methods and Pathology* 2004;84(4):493–501.

328. Schumacher K, Haensch W, Roefzaad C, Schlag PM. Prognostic significance of activated CD8(+) T cell infiltrations within esophageal carcinomas. *Cancer Research* 2001;61(10):3932–6.

329. Zhang L, Conejo-Garcia JR, Katsaros D, Gimotty PA, Massobrio M, Regnani G et al. Intratumoral T cells, recurrence, and survival in epithelial ovarian cancer. *New England Journal of Medicine* 2003;348(3):203–13.

330. Brahmer JR, Tykodi SS, Chow LQ, Hwu WJ, Topalian SL, Hwu P et al. Safety and activity of anti-PD-L1 antibody in patients with advanced cancer. *New England Journal of Medicine* 2012;366(26):2455–65.

331. Hamid O, Robert C, Daud A, Hodi FS, Hwu WJ, Kefford R et al. Safety and tumor responses with lambrolizumab (anti-PD-1) in melanoma. *New England Journal of Medicine* 2013;369(2):134–44.

332. Weber JS, D'Angelo SP, Minor D, Hodi FS, Gutzmer R, Neyns B et al. Nivolumab versus chemotherapy in patients with advanced melanoma who progressed after anti-CTLA-4 treatment (CheckMate 037): A randomised, controlled, open-label, phase 3 trial. *Lancet Oncology* 2015;16(4):375–84.

333. Atkins MB, Lotze MT, Dutcher JP, Fisher RI, Weiss G, Margolin K et al. High-dose recombinant interleukin 2 therapy for patients with metastatic melanoma: Analysis of 270 patients treated between 1985 and 1993. *Journal of Clinical Oncology: Official Journal of the American Society of Clinical Oncology* 1999;17(7):2105–16.

334. Kantoff PW, Schuetz TJ, Blumenstein BA, Glode LM, Bilhartz DL, Wyand M et al. Overall survival analysis of a phase II randomized controlled trial of a poxviral-based PSA-targeted immunotherapy in metastatic castration-resistant prostate cancer. *Journal of Clinical Oncology: Official Journal of the American Society of Clinical Oncology* 2010;28(7):1099–105.

335. Negrier S, Escudier B, Lasset C, Douillard JY, Savary J, Chevreau C et al. Recombinant human interleukin-2, recombinant human interferon alfa-2a, or both in metastatic renal-cell carcinoma. Groupe Francais d'Immunotherapie. *New England Journal of Medicine* 1998; 338(18):1272–8.

336. Askeland EJ, Newton MR, O'Donnell MA, Luo Y. Bladder cancer immunotherapy: BCG and beyond. *Advances in Urology* 2012;2012:181987.

337. Bunimovich-Mendrazitsky S, Shochat E, Stone L. Mathematical model of BCG immunotherapy in superficial bladder cancer. *Bulletin of Mathematical Biology* 2007;69(6):1847–70.

338. Bunimovich-Mendrazitsky S, Halachmi S, Kronik N. Improving Bacillus Calmette-Guerin (BCG) immunotherapy for bladder cancer by adding interleukin 2 (IL-2): A mathematical model. *Mathematical Medicine and Biology: A Journal of the IMA* 2016;33(2):159–88.

339. Kronik N, Kogan Y, Elishmereni M, Halevi-Tobias K, Vuk-Pavlovic S, Agur Z. Predicting outcomes of prostate cancer immunotherapy by personalized mathematical models. *PLoS One* 2010;5(12):e15482.

340. de Pillis LG, Radunskaya AE, Wiseman CL. A validated mathematical model of cell-mediated immune response to tumor growth. *Cancer Research* 2005;65(17):7950–8.

341. Spolski R, Leonard WJ. Interleukin-21: Basic biology and implications for cancer and autoimmunity. *Annual Review of Immunology* 2008;26:57–79.

342. Davis MR, Zhu Z, Hansen DM, Bai Q, Fang Y. The role of IL-21 in immunity and cancer. *Cancer Letters* 2015;358(2):107–14.

343. Pardoll DM. The blockade of immune checkpoints in cancer immunotherapy. *Nature Reviews Cancer* 2012;12(4):252–64.

344. Postow MA, Callahan MK, Wolchok JD. Immune checkpoint blockade in cancer therapy. *Journal of Clinical Oncology: Official Journal of the American Society of Clinical Oncology* 2015;33(17):1974–82.

345. Gettinger SN, Horn L, Gandhi L, Spigel DR, Antonia SJ, Rizvi NA et al. Overall survival and long-term safety of nivolumab (anti-programmed death 1 antibody, BMS-936558, ONO-4538) in patients with previously treated advanced non-small-cell lung cancer. *Journal of Clinical Oncology: Official Journal of the American Society of Clinical Oncology* 2015;33(18):2004–12.

346. McDermott DF, Drake CG, Sznol M, Choueiri TK, Powderly JD, Smith DC et al. Survival, durable response, and long-term safety in patients with previously treated advanced renal cell carcinoma receiving nivolumab. *Journal of Clinical Oncology: Official Journal of the American Society of Clinical Oncology* 2015;33(18):2013–20.

347. Powles T, Eder JP, Fine GD, Braiteh FS, Loriot Y, Cruz C et al. MPDL3280A (anti-PD-L1) treatment leads to clinical activity in metastatic bladder cancer. *Nature* 2014;515(7528):558–62.

348. Wilkie KP, Hahnfeldt P. Tumor-immune dynamics regulated in the microenvironment inform the transient nature of immune-induced tumor dormancy. *Cancer Research* 2013; 73(12):3534–44.

349. Bhat P, Leggatt G, Matthaei KI, Frazer IH. The kinematics of cytotoxic lymphocytes influence their ability to kill target cells. *PLoS One* 2014;9(5):e95248.

350. Boissonnas A, Fetler L, Zeelenberg IS, Hugues S, Amigorena S. In vivo imaging of cytotoxic T cell infiltration and elimination of a solid tumor. *Journal of Experimental Medicine* 2007;204(2):345–56.

351. Carlson JA. Tumor doubling time of cutaneous melanoma and its metastasis. *American Journal of Dermatopathology* 2003;25(4):291–9.

Index

Page numbers followed by f and t indicate figures and tables, respectively.